Green Chemistry Avant La Lettre

The Pine Institute and Resin Chemistry in Aquitaine (1900–1970)

Analysis: Historical Cases in Chemistry

Print ISSN: 3041-0614

Series Editors: Annette Lykknes
(*Norwegian University of Science & Technology, Norway*)
Brigitte Van Tiggelen
(*Science History Institute, USA*)

Published

Vol. 1 *Green Chemistry Avant La Lettre: Pine Institute and Resin Chemistry in Aquitaine (1900–1970)*
by Marcin Krasnodębski

Analysis: Historical Cases in Chemistry - Volume 1

Green Chemistry Avant La Lettre

The Pine Institute and Resin Chemistry in Aquitaine (1900–1970)

Marcin Krasnodębski

Institute for the History of Science, Polish Academy of Science, Poland

World Scientific

NEW JERSEY • LONDON • SINGAPORE • BEIJING • SHANGHAI • HONG KONG • TAIPEI • CHENNAI • TOKYO

Published by

World Scientific Publishing Co. Pte. Ltd.
5 Toh Tuck Link, Singapore 596224
USA office: 27 Warren Street, Suite 401-402, Hackensack, NJ 07601
UK office: 57 Shelton Street, Covent Garden, London WC2H 9HE

British Library Cataloguing-in-Publication Data
A catalogue record for this book is available from the British Library.

Analysis: Historical Cases in Chemistry — Vol. 1
GREEN CHEMISTRY AVANT LA LETTRE
The Pine Institute and Resin Chemistry in Aquitaine (1900–1970)

ISBN 978-981-12-5285-3 (hardcover)
ISBN 978-981-12-5286-0 (ebook for institutions)
ISBN 978-981-12-5287-7 (ebook for individuals)

For any available supplementary material, please visit
https://www.worldscientific.com/worldscibooks/10.1142/12740#t=suppl

Typeset by Stallion Press
Email: enquiries@stallionpress.com

FOREWORD

Analysis: Historical Cases in Chemistry presents up-to-date and informed works on the history of chemical sciences and technologies widely construed to a general learned and scientific curious public interested in the history of science. The series addresses scientists and historians alike, and, in particular, secondary and higher education teachers wishing to use reliable history of chemistry cases in their classrooms. Books from this series stem from recent scholarship in the history of science and include down-to-earth, scientifically and historically accurate cases that can invite reflections on the nature of science or chemistry beyond the case itself.

We are pleased to introduce the first volume of this new book series. It has been a pleasure working with Marcin Krasnodębski in the production of this volume. *Green Chemistry Avant La Lettre: The Pine Institute and Resin Chemistry in Aquitaine (1900–1970)* tells a little-known story of early efforts into Green Chemistry long before the term existed and became so fashionable. Based on extensive research, Krasnodębski unveils the intricate relationships between the local ecosystem, the socio-economical and institutional context, and the material-centred scientific and technical research and development. The book turns a history of resin chemistry rooted in a specific French region into a case study of global history that enables one to reflect on innovation, development, and sustainability practices and strategies. In that sense, there are indeed lessons that can be drawn from history in the frame of our present challenges. The didactic materials to be used in teaching at secondary and tertiary educational levels at the end of the book offer paths in this exploration of past examples of sustainability practices.

Our sincere thanks go to our editor at World Scientific Publishing, Yugarani Thanabalasingam, who came to us several years ago with the idea of a dedicated series and supported us at every step of this journey with trust, patience, and efficiency.

Trondheim/Louvain-la-Neuve
Annette Lykknes and Brigitte Van Tiggelen
April 2024

PREFACE

This book has been in development for many years. Its foundation is the Ph.D. dissertation on the history of the Pine Institute and resin chemistry I wrote from 2013 to 2016 under the supervision of Pascal Duris and Jérôme Pierrel at the University of Bordeaux. The dissertation received a warm reception from numerous scholars and was awarded the prize for the best Ph.D. dissertation of the Division for the History of Science and Technology of the International Union of History and Philosophy of Science and Technology. Since 2016, I have revisited the dissertation's final draft on multiple occasions — amending, correcting, completing, or simply commenting on its content.

The turning point was 2020 when I was awarded a research grant by the Polish National Science Centre to work on the project "Green Chemistry *avant la lettre*: historical and epistemic underpinnings of sustainable practices in chemistry and the chemical industry". During the project, I delved into the complex history of green chemistry spanning several decades, and came to see the history of resin chemistry in a new light as a predecessor to what is heralded as a revolution in chemical sciences today.

What was initially meant to be a simple translation of my Ph.D. work into English, therefore, turned into a much more comprehensive overhaul of the initial text, involving the study of previously unexplored archives and new literature. Most importantly, it entailed rethinking the very content of every chapter by adding novel elements and streamlining the core narrative. As such, while the "skeleton" of the book still resembles the original dissertation, its content moves towards previously unexplored territories.

Since the book uses numerous unpublished archival sources, all the translations of the French documents are my own unless indicated otherwise.

All this work would not be possible without the help of the series' editors, Annette Lykknes and Brigitte Van Tiggelen, who undertook the arduous and ungrateful task of reviewing the subsequent iterations of the book's multiple drafts and who provided me with invaluable feedback. I want to also thank anonymous reviewers who, while critical, helped me to better frame my story. The writing of the book was also possible thanks to the supportive work environment of the Institute for the History of Science of the Polish Academy of Science and, of course, to the already mentioned funding from the National Science Centre that gave me extraordinary freedom of research, so rare and precious in the contemporary academia.

Finally, I want to thank my wife, a chemist passionate about sustainability, for sharing her insights and suggestions.

Funding Acknowledgement

The writing of this book was funded as a part of the project "Green Chemistry *avant la lettre*: historical and epistemic underpinnings of sustainable practices in chemistry and the chemical industry", carried out as the Sonata 16 grant no. 2019/35/D/HS3/00614 of the Polish National Science Centre.

CONTENTS

LIST OF FIGURES

LIST OF TABLES

INTRODUCTION

The concept of green and sustainable chemistry emerged organically in Europe and the United States (US) during the 1990s [Krasnodębski, 2022; Linthorst, 2023]. Green chemistry serves as an umbrella term, encompassing a wide range of ideas that integrates environmental and sustainability-related considerations into the practices of chemical laboratories and factories. Green chemists study a diverse array of subjects, including toxicity, waste reduction, laboratory safety, and, notably, the use of renewable raw materials.

This novel concept has gained widespread acceptance worldwide, with many universities now offering programs in the field. Numerous journals, such as *Green Chemistry*, *Green Chemistry Letters and Reviews*, *Trends in Green Chemistry*, *Current Opinion in Green and Sustainable Chemistry*, and many more, contribute to the ongoing discourse on the frontiers of greenness. Dozens of textbooks attempt to define the notion of green and sustainable chemistry while describing its methods and promising research trajectories [Antoniotti, 2013; Torok and Dransfield, 2017; Beach and Kundu, 2017]. One of the most influential definitions of the discipline was formulated within *Green Chemistry: Theory and Practice* (1998), which laid down the so-called 12 principles of green chemistry [Anastas and Warner, 1998].

Despite some critical studies pointing out the limitations of the green chemistry framework [Roberts, 2006; Nieddu *et al.*, 2014], the general community of chemists lauds it as a new paradigm aptly responding to the environmental crisis. The importance of sustainable chemistry was notably recognised in 2005 by the Nobel Prize committee when it awarded the prize in chemistry to Yves Chauvin, Robert H. Grubbs, and Richard R. Schrock for the development of the metathesis method in organic synthesis. The committee described their work as "a great step forward for 'green chemistry,' reducing potentially hazardous waste through smarter production" [Nobel Prize Committee, 2005].

However, one may wonder whether some of the questions raised by the practitioners and proponents of green chemistry are really that novel. Were the chemists in the past not preoccupied with the scarcity of raw materials or with over-dependence on petroleum? Green chemistry is often presented in the literature as a new opening, a radical departure from the old, obsolete, and polluting chemical industry, uninterested in environmental issues [Woodhouse and Breyman, 2005]. But while the formalisation of the concepts of green and sustainable chemistry is a recent invention, the actual history behind them stretches back to the nineteenth century and touches not only upon the history of chemistry but also the environmental history and the origins and development of modern capitalism.

While historians of science and technology have extensively researched the history of biomass use as a fuel [Ekerholm, 2012; Manuel, 2021; Krasnodębski, 2021], early attempts in what is now known as sustainable chemistry often remain overlooked. One of the best-studied examples is chemurgy, a scientific movement active in the US during the 1930s and 1940s, with the goal of making crops the primary raw material for the chemical industry [Finlay, 1990, 2004; Uekötter, 2021]. However, this initiative was far from being unique. Why is there a lack of literature on the history of disciplines, frameworks and concepts that focused on the study of renewable materials in the first half of the twentieth century?

The first reason is that they were overshadowed by the tremendous success of fossil fuels. Non-renewable resources, such as coal and later petroleum, were at the heart of organic chemistry in the middle of the nineteenth century and constituted the raw materials par excellence of the Second Industrial Revolution. It is a truism that petroleum became, by the middle of the twentieth century, the lifeblood of the global economy by providing fuels, solvents, plastics, and many other derivatives. However, because of the importance of fossil hydrocarbons, their history often stole the spotlight from other substances that also fascinated organic chemists over a century ago. The virtues of renewable biomass were extolled long before the term became popularised in the 1970s, in reaction to the oil crisis.

The second reason is due to the context of the emergence of these various projects. In fact, in the interwar period, the primacy of petroleum was still far from established. Many dreamt big about a future where nations would rely on their own renewable raw materials and where economies would expand without depending on hydrocarbon deposits that were believed to last only a few decades more. These aspirations were often coloured by autarky: a nationalistic obsession of self-sufficiency preached in a world where national and military interests were at the centre of many industrial and scientific endeavours. This situation led some historians to see these biomass valorisation projects through the lenses of failed ideological aspirations of the period. But these early attempts in creating expertise on renewable materials and sources of energy were much more than expressions of ill-conceived nationalism. They stemmed from a genuine concern about the long-term economic development as well as from the local aspirations of people working in smaller industries who placed their hope in scientific progress that was meant to bring them prosperity and economic emancipation.

One of the industries that challenged the growing primacy of fossil hydrocarbons in the early twentieth century was pine forestry. Pines are hardy trees that can grow in both dry and moist soil, and are widespread in many different climates. Their resinous wood has been used since antiquity for manufacturing so-called "naval stores", tars and pitches necessary for building and maintaining wooden sailing ships. In a way, pine was one of the most universal raw materials known throughout huge parts of human history. Its strategic value put this coniferous tree at the heart of economic, political and military conflicts between European nations in the early modern period [Outland, 2004, p. 11].

Pine's versatility is due to resin, colloquially called gum, a sticky, oily substance easily recognisable for its particular odour. The tree, when wounded, secretes resin that can be collected for many days with the use of appropriate equipment. Pine gum was believed to have medicinal properties, but it was also used as a substitute for tar and for the manufacture of torches. It was not, however, until the eighteenth century that resin was given a new life thanks to the spread of distilleries. Through distillation, resin was transformed into two components: volatile turpentine and solid rosin. Turpentine became particularly in demand on the global markets throughout the nineteenth century because of its applications as a lamp fuel (before kerosene), and as a solvent for paints and varnishes.

By the 1920s, pine forests held more promise than ever before, thanks to the progress of science. Pine charcoal gained sudden attention as a potential fuel for vehicle gasifiers; turpentine was not only a solvent anymore but turned into an essential raw material for the synthetic camphor industry and was poised to be used in the manufacture of other synthetic materials and fuels; rosin, long considered a wasteful byproduct of resin distillation, was coveted in the manufacture of paints and varnishes, soaps, adhesives, road coatings as well as in paper production. This new thriving industry developed alongside a flourishing scientific discipline: pine resin chemistry.

While pine forests were planted and exploited in many different countries, two nations dominated the global production of pine resin derivatives in the first decades of the twentieth century, accounting for 90 per cent of the total supply: the US (70 per cent), thanks to its vast natural forests in the southern states, and France (around 20 per cent), thanks to its giant artificial forest planted specifically for this purpose in the Landes of Gascony in Aquitaine. These two industries were assisted by two institutions: the Pine Institute (*Institut du Pin*), located in Bordeaux, that laid the foundations for resin chemistry before World War II, and the Olustee Station, in Florida, whose golden age started after 1945. These organisations not only conducted chemical research but engaged in a comprehensive reflection on the technological, industrial and economic dynamics of resin-based materials. Both of them were established in the interwar period, but by the end of the 1960s, the Olustee Station had disappeared while the Pine Institute was completely transformed.

This book aims to explore the history of pine resin chemistry — a discipline that, long before the modern green chemistry conceptualisations, attempted to address similar issues. The interest in this discipline stems not only from the fact that pine resin is a renewable resource with ample supplies that can emancipate various sectors of the economy from the oil-dependent supply chain, but also from the fact that its practitioners incorporated sustainability-oriented considerations into their work. Its institutions were deeply rooted in local communities and provided a comprehensive approach to the challenges of building an alternative self-sufficient economic model, mindful of the interests of the local populations. Out of the two institutions mentioned before, the Pine Institute, and therefore the French resin chemistry, is at the heart of this study, specifically due to the fact that it fully exemplifies the characteristics that made resin chemistry a forerunner of modern green chemistry. However, this book also examines the American case and explains why it never fully replicated a similar set of values and institutional settings.

This research project builds upon existing scholarship in the history of science. The case study contributes, in particular, to a growing body of literature on the history of the French science "in the provinces" outside major Parisian research centres that dominated the perception of the French research landscape until very recently [Paul, 1985; Nye, 1986]. One of the reasons why French resin chemistry and many other small scale research endeavours in France remained long understudied by historians of science was due to the lack of interest in small provincial faculties of sciences in a country dominated by the universities in Paris or state-owned engineering schools.

Our focus is Bordeaux in Aquitaine, a city known to many as the wine capital of France. But there was a time when the city was called the capital of "wine and pine" (*capitale du vin et du pin* in French). Located on the banks of the river Garonne, Bordeaux sits precisely at the boundary between vast vineyards and a massive pine forest. The forest covers almost the entire territory of the department of the Landes and a huge part of the department of Gironde. The traditional name of the region is the Landes of Gascony, and it used to be a scarcely inhabited moor before the massive forestation projects in the nineteenth century. It was the Bordeaux Faculty of Sciences whose laboratories started to study resin chemistry in 1900 to help the local resin industry develop new products from this tremendous natural resource. In this sense, this study joins a growing, although still very inadequate, cohort of works specifically devoted to the history of the Bordeaux Faculty of Sciences [Paul, 1996; Lascombe, 2009; Maison, 2013].

At the same time, by studying the history of a scientific discipline (resin chemistry), the book aligns with other recent works exploring various subdisciplines of chemistry, such as solid-state chemistry [Teissier, 2014], electrochemistry [Petit, 2013] and catalysis [Voillequin, 2010]. More and more historians are concentrating on more circumscribed fields of study, showing how chemistry, or any branch of science, is, in fact, a bundle of individual trajectories entrenched in their socio-economic and intellectual environments, pointing out the historically contingent character of science [Shapin, 2010].

Readers familiar with the sociology of science and technology may also find that the case study corresponds to frameworks elaborated in this discipline to describe complex relations between academia and the industry. On the one hand, a resin-producing forest can be considered a "local production system" or an "industrial district" [Crouch *et al.*, 2001; Daumas, 2007]. These two concepts help to understand the microcosm of relations between various stakeholders (companies, labour unions, public authorities, universities) on the local level. This includes the transmission of know-how, establishment of trade customs and standards, construction of infrastructure, and research and knowledge production.

Others may find the case study to fit the framework known as the "Triple Helix" model, which relates to the relations between three interdependent entities: the state, university and industry [Shinn, 2002; Ranga and Etzkowitz, 2013]. The concept is both descriptive (it marks the departure from the traditional binary logics: State-Industry, State-University) and normative (scholars try to find the optimal way to systematise the collaboration between these entities). The case study of the Pine Institute indeed provides an interesting illustration of how such close collaboration looked like across decades.

These bibliographical and conceptual references may help contextualise the basic premise of the work and understand the considerations it stems from. However, those who are not familiar with them can rest assured: this is a self-contained study that guides the readers through the challenges of constructing a coherent historical narrative. No prior knowledge in history or sociology of science or in chemistry is required to follow the argument.

Highlighting the variety of perspectives is nevertheless crucial for understanding that there is no single and unequivocal way of telling a story. By bringing them together, the aim is to hint at the heterogeneity of questions and problems raised by a single case study but also to underline that these perspectives are often dependent on one another. This book has two different but inseparable study topics: a scientific discipline (resin chemistry) and a research institution (the Pine Institute). By discipline, I mean a certain organised body of knowledge as well as novel research questions generated by it. But a "living" discipline needs a physical substrate: researchers, journals and institutions devoted to it. While some disciplines can develop through invisible colleges of scientists collaborating in universities around the world, the case of resin chemistry is the opposite. Pine resins were, of course, studied by chemists in different countries, and a tremendous amount of literature from the late nineteenth and early twentieth century was produced in Germany, including some foundational books in the field. However, it was in France, in Bordeaux, where an institution fully dedicated to pine resin chemistry was established, and that was the discipline's driving force over the decades. Resin chemistry existed because the Pine Institute of Bordeaux reached a critical mass of scholars, expertise and funding. By describing the values, views, contributions, and trajectories shared by the scientists working for the Institute, we can understand the origins and frontiers of resin chemistry as a full-fledged discipline.

The twofold approach; that of the discipline and the institution, permeates the study as a whole. This book takes the view that the relationship between the discipline and the institution should be analysed in a symmetrical way. On the one hand, the book explores the history of a research institute with its staff, directors, and research topics, as well as administrative quirks, to understand the history of a discipline. On the other hand, it focuses on a discipline and the questions it raises in order to understand why the Institute looked the way it did. Both approaches are necessary for this case study to be useful in thinking about issues surrounding sustainability today. Sustainable chemistry is not about building a body of knowledge on renewable raw materials and energies but also about organically integrating these perspectives through research institutions into local economic settings with all their individual and unique challenges. This is the main point the book attempts to convey.

A word of caution is necessary at this stage. This book does not study the economic history of the resin industry, nor does it try to cover the complexity of the international resin trade over decades. Such an industrial history of resin certainly would be a valuable companion to this book, but here, I chose to focus specifically on resin science as a discipline in its social, political, and institutional context. It does not mean that the book ignores the economic history. Readers will find in this work rich and detailed information about the production of resin and its derivatives in different countries and especially about the major tendencies transforming the industry in the first half of the twentieth century. However, this information is used only to justify and support arguments concerning the history of resin-related research itself and to provide a richer and more complete overview of the background in which resin chemistry developed. This philosophy extends over many other subjects that are part of the history of resins in a broad sense: national policies concerning pine forest use (especially from an environmental perspective) and the intricacies of knowledge transfer between the laboratory and the factory, not to mention the history of scientific results concerning the composition of various types of resin. All of these are explored throughout the book as accessory topics, but they are not the main focus *per se*. In a certain way, I hope to initiate through this work a broader discussion on these various aspects and inspire more research on the history of techno-scientific alternatives to fossil feedstocks.

The book is divided into six chapters, each highlighting some crucial characteristics of resin chemistry as a unique discipline. Chapter 1 introduces the Pine Institute and presents its history from inception until World War II. The chapter's purpose is not only to set the stage for further discussions and present key stakeholders but also to explore the Institute's unusual and multilayered character as a hybrid research unit, bringing together a wide array of stakeholders. Chapter 2 is a historical and somewhat epistemological introduction to resin chemistry, highlighting some of its recurrent challenges. This chapter's goal is, above all, to insist on the interminglement between academic and industrial considerations and priorities in the context of the pine forest uses. Chapter 3 integrates the two previous approaches to offer an overview of the history of the Pine Institute and the French pine resin chemistry during and after World War II, putting particular emphasis on the collaborations between the Institute and the local industry. It shows that industrial, economic, and political constraints and decisions can shape the direction of research and, ultimately, the fate of an entire discipline. Chapter 4 retells the Pine Institute's history from a financial and economic point of view. It explores the intricate relationship between economic cycles and the prosperity of scientific institutions that "sell" promises of progress to their stakeholders. In Chapter 5, we look more in detail at the American research landscape in resin chemistry. The American case is interesting because it not only demonstrates the influence that French science had abroad through various intellectual exchanges but also because it reveals an alternative model of institutional evolution. More interestingly, it shows that while resin chemistry developed all over the world, its French roots remain crucial for understanding its particular character. Chapter 6 is a detailed case study of the standardisation of resinous products, highlighting how scientific definitions can become tools of business pressure in the hands of local production systems. It also points out the inherent messiness of terminology in the field. Some of these chapters (notably Chapters 5 and 6, but also, to an extent, Chapters 1, 2 and 4) can be read as self-containing stories. Of course, many interconnections will make sense only if the book is read in its entirety, but readers are invited to read chapters in any order, depending on their interests.

Finally, after the general conclusions, there is an annex specifically devoted to teachers interested in using the book in their classes. The annex constructs a lesson grid for students at different levels of education. Its goal is to present the key takeaways from the case study on issues such as sustainability, the history of chemistry, and the history of science in general. The lesson grid can also act as a compass for those who prefer a more problem-driven reading of the book and want to focus on some of its particular aspects.

Together, regardless of the reading strategy adopted, all chapters and annexes constitute a comprehensive guide to one of the most intriguing but overlooked scientific endeavours of the last century.

Figure 1: The Landes pine forest.

Source: Cabaussel, 2006 — Larrousiney, CC BY-SA 2.5. https://fr.wikipedia.org/wiki/For%C3%AAt_des_Landes#/media/Fichier:LandesDeGascogne.png

Chapter 1

SETTING THE STAGE: ORIGINS AND FIRST YEARS OF THE PINE INSTITUTE (1900–1939)

> "Do institutions make people? Or: Do people make institutions? Of course, the correct answer to both questions is yes."
> [Tool, 1993, p. 254]

The study of scientific organisations is, perhaps scientific biographies aside, one of the most classic approaches in the discipline of the history of science. Historians of science passionately explore the intricacies of entities such as learned societies [Hahn, 1993; Seitz, 2007; Tinniswood, 2019], institutes of industrial research [Paul, 1980; Fonteneau, 2010; Krasnodębski, 2018], and different types of laboratories [Pestre, 1990; Johnson and Schaffer, 1994; Holl, 1997]. Concentrating on an organisation usually permits scholars to easily and quickly delineate the perimeter of their objects of study. They can focus, from then on, on the organisation's members, key stakeholders, collaborations, and the most important lines of research. Through institutional histories, researchers can construct more sophisticated and less tangible conceptual categories, such as disciplines, research schools, research communities, and many others [Servos, 1993; Nye, 1993; Reinhardt, 2001; Teissier, 2014].

However, studying the history of organisations presents a unique set of challenges. New institutions often emerge from previously existing ones and evolve through time. It is notoriously difficult to pin down their origin. Some organisations can keep the same name over decades, if not centuries, but drastically change their research profile and the nature of their activities [Kaiserfeld, 2013]. The problem of determining the founding date arises in the case of the Pine Institute in Bordeaux. For example, in 1987, the Institute celebrated its 50th anniversary.[1] A beautifully illustrated brochure published on this occasion explained that the Institute was created on 27 October 1937, when a ministerial decree confirmed the creation of the Pine Institute of the Faculty of Sciences of Bordeaux. The year 1937 was also retained in the memories of former directors from the 1990s and 2000s [Krasnodębski, 2014]. However, after a closer examination of the Institute's publications and archives, it is necessary to revise the founding date. In fact, an internal document outlining the history of the Institute from 1996 indicates that 1937 was not "the birth, but the christening of an already adult structure".[2] Furthermore, there was already a review called *Bulletin de l'Institut du Pin* (Pine Institute Bulletin) between 1924 and 1939. A reading of its first issues points out the existence of something called "l'Association Institut du Pin", or the Pine Institute Association, that had been established in 1922. Did this Association transform into a chemical institute? In that case, why was its history forgotten afterwards? Further questions arise when we delve deeper into the institution's older archives; for example, there is an application for funding written by the Institute's long-time director, Georges Brus, in 1965, in which he claimed that his laboratory was no less than 65 years, placing its beginning in 1900.[3]

---

[1] Departmental Archives of Gironde, Pine Institute Archives, box "Divers". "50ème anniversaire de l'Institut du Pin", 1987, p. 5.

[2] Departmental Archives of Gironde, Pine Institute Archives. R. Bentejac, "Historique de l'Institut du Pin et de ses activités dans le domaine de la chimie", 1996, p. 4.

[3] Departmental Archives of Gironde, Pine Institute Archives, file "Institut du Pin, Ve Plan". "Recherche universitaire hautement spécialisée, construction et équipement de l'Institut du Pin", 25 February 1965.

1900, 1922 or 1937? When was the Pine Institute created? This chapter tries to disentangle this knot to understand, decade by decade, what the Pine Institute was before World War II. The chapter begins with an overview of the social, economic and environmental factors that made the rise of resin chemistry possible in Bordeaux. Then it explores the early attempts to institutionalise it in the 1910s, 1920s and the 1930s and, in the following sections, it presents the directors, staff and various non-research activities of the Pine Institute in its early days, as well as its consecutive administrative transformations. In the final section, the chapter discusses the history of the Institute through the lens of its internal review.

1. The Landes Forest in the Nineteenth Century: An Economic Prehistory

The Landes Forest is today the largest French forest massif, but this was not always the case. For hundreds of years, the Landes of Gascony was an arid desert during summer and a pestilent swamp in winter. The region was inhabited by shepherds, considered to be half-savage by some French intellectuals.[4] Small forest islands were scattered across its territory and enormous-moving dunes along the coastline separated it from the Atlantic Ocean. These piney woods became important for the shipyards of Bordeaux as far back as in the seventeenth century, but the supply was sufficient, and no forestation projects were pondered over the period. The idea of forestation of the region was put forward much later, only at the end of the eighteenth century. One of the first attempts of this sort was conducted by the Swiss engineer, Daniel Nézer who wanted to reclaim the Landes wasteland by building canals (to dry the land) and a route network and then by cultivating wheat, oat, rye, and vine, and by extending pine plantations. His project, which started in 1765, was an utter failure, and Nézer died in misery. Out of the various

[4] Some writers considered the local inhabitants to be "closer to apes than to men" due to their primitive living conditions. See [Sargos, 1997, p. 61].

cultures, only the pine plantations survived him. Nézer's experience, however, inspired local nobility to experiment with pine plantations, but their place in the local economy was still far from established [Sargos, 1949; Sargos, 1997; Krasnodębski, 2019]. Many different initiatives for valorising the Landes followed in the nineteenth century, exploring alternatives, such as clover, beetroot, different vegetables, peanuts, rice, various species of animals (Swiss cows, English pigs, Tibetan goats) as well as other trees, notably oaks.

Nonetheless, pines continued to affirm their place in the local landscape, thanks to their use in the fixation of the coastal dunes. Leaders of the communities neighbouring the dunes felt threatened by these moving mountains, which had been accused of having submerged entire villages. Although we know these fears were largely spurious, as the dunes moved to and fro in a very limited geographical area, many assumed during the eighteenth century that these dunes would venture further inland and threaten bigger cities with sand, the famous port of Bordeaux included. As a consequence, these dunes had to be stopped and "fixed" in place. What technology could achieve such an ambitious task? After many attempts, the most successful operation for sand dune stabilisation was conducted by an engineer named Nicolas Brémontier, who used pine trees for this goal. Between the 1790s and 1870s, 88,000 hectares of dunes were covered with a dense pine forest that profoundly transformed the coastal landscape [Maizeret, 2005]. Brémontier's success contributed to an improvement in the perception of pines, trees powerful enough to reshape the territory and tame the forces of nature. As a consequence, pines became increasingly included in local development plans as a favoured means of transforming wasteland into valuable terrain. The pinnacle of this tendency was the enactment of a special law, issued in 1857 with the explicit endorsement of Emperor Napoleon III, which obligated the local communities to plant pine trees to improve the sanitation of the Landes moors.

Pine planting supporters combined patriotic rhetoric (economic independence from foreign wood and resin importation) with proto-ecological considerations (enhancement of the local climate) [Moriniaux, 1999, pp. 131–134]. The enthusiasm over the promises of progress was not shared by everyone, though. The massive forestation would disturb the local agropastoral equilibrium. Contrary to the dismissive perceptions of the wealthy bourgeoisie and government officials, the Landes of Gascony was not a "wasteland" for its inhabitants. Flocks of sheep were a source of milk, wool, and, most importantly, manure, which fertilised the soil and allowed the cultivation of rye and millet for manufacturing textiles. Contrary to the traditional (mis)representation that the local population was poor, the local communities were fully self-sufficient and could successfully trade with the outside world [Sargos, 1997, p. 77]. Due to the paucity of plant life, however, sheep needed vast spaces to feed themselves. As a consequence, the extension of woodland was considered a threat to the traditional lifestyle. Local farmers and shepherds refused to give away the pastures used by their animals, and they even notoriously set the new pine forests on fire. The arsonists were protecting their revenues and economic self-subsistence against the interests of the slowly growing pine resin industry and powerful landlords extending their property. Arson became one of the most prosecuted crimes in the second half of the nineteenth century in the Landes.

The situation changed drastically in 1861 after the newly elected American president Abraham Lincoln ordered the Continental Blockade during the first days of the American Civil War. North Carolina, South Carolina, and Georgia had for years provided both the emerging German chemical industry and the British Navy with resinous products [Outland, 2004]. The blockade accelerated the reforms that had been initiated in the 1850s by Napoleon III in France. The inhabitants of Aquitaine wanted to profit from the exorbitant prices that resinous products yielded and became involved more willingly in the management of the existing forests and the plantation of new trees. In parallel, local communities sold off massive swaths of their land to huge forest investors. Charles Prat, a labour union leader, explained many years later that the American Civil War had led to "frenetic plantations" all over the region [Prat, 1947b]. While the pine forest in Landes and Gironde covered only 100,000 hectares at the end of the eighteenth century, the surface of the *forêt landaise* had, by the beginning of the twentieth century, grown ten-fold, up to almost one million hectares, and the resin industry became the driving force of the local economy [Sargos, 1997, p. 501]. A new "local production system" was formed with a complex network of relations between distillation plants, forest owners, forest workers, and distributors.

However, the fruits of this transformation were bitter. After the end of the Civil War, intercontinental trade resumed. Prices fell dramatically, the promise of prosperity did not materialise, and the whole region suffered a severe economic crisis. For example, in 1864, the price of a barrel of 225 litres of gum reached 300 francs. Five years later, in 1869, it fell to 45 francs [Sargos, 1997, p. 501]. Unfortunately for the local population, it was too late to turn back after the war and give up the reforms. The property structure in Landes and Gironde had changed, and land owners continued the forestation projects, hoping to profit from pine resin in the long run. The Landais and Girondins, struggling to survive, had no choice but to accept new forest jobs and hope that prices would increase in the global markets. Over the course of ten years, the prosperity of previously isolated agricultural communities became dependent on turpentine prices on major resin markets in London, Wilmington and Savannah.

The gravity of the economic crisis in the 1870s was acknowledged by local politicians who tried to find a way to reaffirm the position of the Landes products on the global market despite the American competition. In a fiery speech in 1872, Alexandre Leon, elected regional councillor of Gironde, made a passionate plea for scientists from Bordeaux to help improve the colour of resinous products. Colour was one of the main factors affecting the price of the products; the lighter, the better. The call remained unanswered, but the local chemists were not alone to blame for this lack of reaction. The Faculty of Sciences of Bordeaux simply lacked financial and material resources to establish more robust collaboration with the industry [Paul, 1985, p. 201].

The resin industry remained unstable in the following decades; years of prosperity interspersed with years of economic difficulties.

By the end of the nineteenth century, the region was hit by another crisis, and the Landes authorities appealed to the Bordeaux Faculty of Sciences once more. Pierre Duhem, famous physicist and historian of science from Bordeaux, related this event in his notes: "At the session of April 1899, the General Council of Landes expressed the wish that the Bordeaux Faculty of Sciences become interested in the resin industry and educate chemists capable of improving it." [Duhem, 1901, p. 3] This time, the academics were more responsive. In 1899, Dean Georges Brunel asked Maurice Vèzes, a specialist in inorganic chemistry, to explore the topic and help the resin industry.

2. Laboratory of Chemistry Applied to the Resin Industry (1900–1921)

The *Laboratoire de chimie appliquée à l'industrie des résines* (Laboratory of Chemistry applied to the Resin Industry) was the first of its kind in the world and laid the foundations for the entire discipline of resin chemistry. It was the heart of what would become the Pine Institute in the following years. To understand the environment in which the Pine Institute was born, it is therefore important to explore the circumstances that made the establishment of the laboratory possible and the reasons behind its fall from grace.

Maurice Vèzes — An Unlikely Creator of Resin Chemistry

Maurice Vèzes (Figure 2) graduated from one of the most prestigious French schools — *École normale supérieure* — and, between 1888 and 1893, occupied the position of teaching assistant at the same institution. In 1894, he moved to Bordeaux and in 1901, he became the chair of inorganic chemistry at the Faculty of Sciences of Bordeaux [Dupont, 1935]. He held many positions inside the academic world throughout his life, but he is most remembered today for having established in 1900 the Laboratory of Chemistry Applied to the Resin Industry, of which he remained the director until 1925.

Figure 2: Maurice Vèzes (1864–1935), creator of the French resin chemistry and director of the first resin laboratory in Bordeaux.

Source: *Bulletin de l'Institut du Pin*, 2, 1935, p. 22.

In 1935, after the death of Vèzes, his contemporaries heaped praise upon his involvement in the resin industry: "[His previous work] was not enough for him; ... he thought that it was his duty, with the knowledge he had, to be more useful by serving a great local industry deprived of an enlightened leadership before." [Chaine, 1935] Duhem's testimony from 1901 remains much more nuanced, especially concerning Vèzes' initial reaction to the call: "He does not want at all to abandon his research in inorganic chemistry which he is so passionate about. He does not know the resin industry. The dean insists; a service is to be rendered that cannot be deferred." [Duhem, 1901, p. 3]

Vèzes' reservation is understandable, given the challenge of studying a new broad topic without prior experience. When Dean Brunel entrusted Vèzes with the task of developing resin chemistry research and teaching in 1899, his first step was to visit local workshops and consult forest workers to understand the practical problems faced by the industry. In 1899, Vèzes organised five lessons on resin chemistry that drew a very wide audience involved in the local industry. In 1900, ten more lessons followed, but he realised that what was needed was far more than education about chemical substances and processes; a fully equipped chemical laboratory was necessary to conduct further research on new products and techniques. He explained that teaching was not sufficient because what the local industrial leaders wanted was not to "to study the past and present knowledge, but to prepare the future" [Vèzes, 1934, p. 6].

Interestingly, in 1900, in his first year fully devoted to resin chemistry, Vèzes published only one scientific article, "Un perfectionnement dans la fabrication des huiles de résine" (On Perfecting Oil Resin Manufacture), which was more of a summary of different inventions than his own laboratory research, but he published over a dozen articles on the state of the industry in France, Europe, and the United States (US). The US remained the largest manufacturer of resin materials in the world, and Vèzes wanted to familiarise the French producers with the economic situation of their main competitor. Vèzes believed that science and economic activities were interrelated and to be discussed concurrently, which appears to be the most remarkable feature of his philosophy.

In this respect, he continued the tradition of other important scientific figures in Bordeaux. From a theoretical standpoint, the necessary link between science and industry was described by the previously mentioned physicist Duhem. While Duhem is today mostly known for his contributions to the history and philosophy of science, he was passionate about industrial applications of chemistry and deplored the Faculty of Sciences of Bordeaux for building itself a "palace" instead of a "workshop" in 1885.[5] For him, the chemical laboratories should have been, in fact, semi-industrial plants closely collaborating with local industry leaders. In 1899, Duhem proclaimed: "The factory would save the laboratory from byzantine sterility; the laboratory would protect the factory against empiricism and routine." [Duhem, 1899, p. 386]

But Vèzes also had more practically oriented predecessors who paved the way for the establishment of his laboratory. After the shock following the defeat in the Franco-Prussian war in 1871, many blamed the gap that supposedly grew in France between science and industry [Smith, 1979, p. 307–312]. Under the Third Republic, numerous new industry-oriented engineering schools were brought to life, such as *École supérieure de physique et de chimie* (Higher School of Physics and Chemistry) created in 1885 and *École supérieure d'électricité* (Higher School of Electricity) established in 1894. In Bordeaux, it was the student of the French chemist Louis Pasteur, Ulysse Gayon, who was instrumental in bringing the industry closer to the Faculty of Sciences. In 1880, he established the viticulture research station, which is today the world-famous wine-research centre, the Faculty of Oenology. In 1891, he created the *École de chimie appliquée à l'industrie et à l'agriculture* (School of Chemistry Applied to Industry and Agriculture) that became later the Graduate School of Chemistry, Biology and Physics of Bordeaux. Vèzes cited Gayon explicitly as an inspiration for his project, alongside the Brewery School of Nancy, created in 1893 to reinvigorate the local beer industry [Vèzes, 1905, p. 3].

Rise and Decline of the Resin Chemistry Laboratory

The new laboratory required an annual budget of approximately 7,000 francs, which, according to Vèzes, the Faculty would not be able to cover [Vèzes, 1934]. He believed that it was up to public

[5] The shift in the vocabulary is noteworthy as only a few decades before, the construction of "chemical palaces" was cherished in Germany. See [Morris, 2015, p. 141].

authorities, such as the general councils of Gironde and Landes or the chambers of commerce of Bordeaux, Mont-de-Marsan and Bayonne, as well as up to local industry leaders to fund the initiative. In November 1900, the amount raised was 4,500 francs, thanks to the publicity in local newspapers. Despite these relatively modest resources, the Laboratory of Chemistry Applied to the Resin Industry was inaugurated the same year at the Faculty of Sciences of Bordeaux.

The resin laboratory filled three distinct roles. It was (1) a technical teaching laboratory training engineers for the resin industry, (2) a laboratory of analyses and technical tests ensuring the quality of resinous products, and finally, (3) a research laboratory solving the practical challenges faced by the local industry and developing new products. The research role was considered particularly important. The laboratory's director explained that research:

> [F]ocuses on the points indicated by the industry leaders themselves, on the questions that were formulated in the practice [of the industry] .... It is in this aspect, in which the fundamental character of this enterprise can manifest itself: this laboratory is a laboratory of applied chemistry, a type of test plant, and not a pure chemistry laboratory; we will not work on theoretical problems there, we will do pure science only insofar as it is needed to provide a solid basis for the applications [Vèzes, 1905, p. 3].

Vèzes did not give up on teaching though. Between 1899 and 1906, he gave lectures every year on a topic, which reflected his own progression in the field (Table 1).

There were years when Vèzes emphasised broader theoretical problems, such as the history of studies on the chemical composition of rosins. In others, he explored in detail the much more mundane technical challenges of resin harvesting. Inside a single disciplinary matrix, he combined the know-how of a chemist with that of a forest worker.

While the classes themselves certainly answered the needs of the industry, their formula, periodic lectures in Bordeaux, was not well

Table 1: Classes given by Maurice Vèzes in 1899–1906[6]

Year	Topic
1899/1900	Resin harvest and distillation
1900/1901	Chemical history of dry resin products (rosins)
1901/1902	Pyrogenated rosin derivatives (resin oils)
1902/1903	Resin harvest and distillation
1903/1904	(No lectures because of Vèzes' health problems)
1904/1905	Soaps and fatty acids
1905/1906	Analytic methods applicable to fats and resins

adapted. The industry technicians and engineers who worked in the workshops spread throughout the vast region of the Landes of Gascony simply could not attend their lectures regularly. From 1907/1908, the organisation changed [Vèzes, 1908]. Instead of lectures, Vèzes invited select candidates to his laboratory for a few days of intensive practical internships on fraud detection practices as well as on other problems concerning the improvement of the quality of resinous products. At first, the classes gained immense popularity among industry workers, even though they were not free of charge. There were so many candidates that a waiting list had to be established. They were divided into different categories depending on their financial contributions to the laboratory. And yet, whatever the initial successes were, the laboratory's activity report from 1911/1912 (the only one from the decade 1910–1920 that has survived until today) does not mention teaching at all. It may be due to the fact that, over the course of a few years, those interested in learning successfully applied and got the education they wanted. But this absence may also indicate a slow decline of the laboratory. This can be confirmed by the fact that Vèzes and his laboratory became much less scientifically productive in the years 1911–1920 in comparison with the years 1900–1910 (Figure 3).

[6]Table established from data in M. Vèzes, Les rapports du fonctionnement du laboratoire de chimie appliquée à l'industrie des résines, 1901–1908.

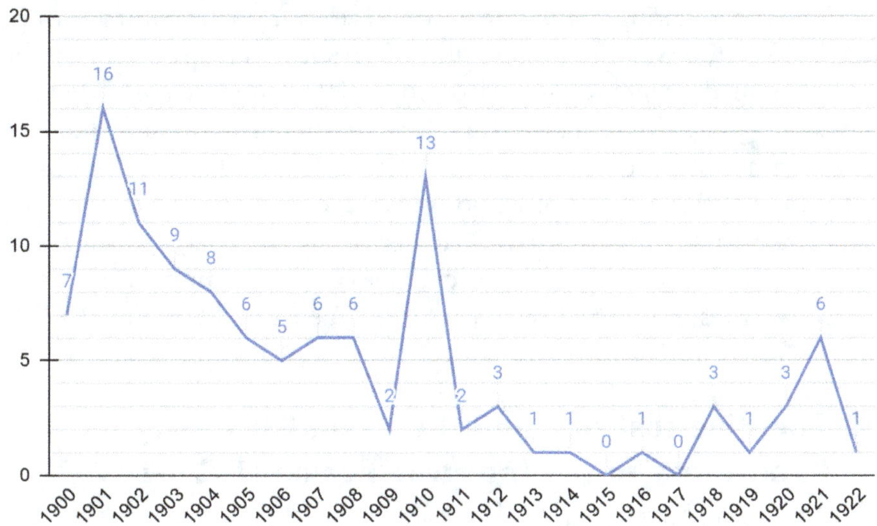

Figure 3: Number of articles published by Maurice Vèzes per year.
Source: *Bulletin de l'Institut du Pin*, 2, 1935, pp. 30–39.

Vèzes published, on average, eight articles a year in the first decade of the twentieth century but only 1.5 in the subsequent decade. This can be partly explained by the growing complexity of the problems studied. Resin chemistry was developing as a discipline, and more and more articles, instead of formalising the know-how already present in the industry, had to proceed from original laboratory research. And yet, it is equally possible that his laboratory faced financial difficulties. After the first prosperous decade throughout the 1910s, the scientific production of the laboratory dwindled alongside its budgets. In Chapter 4, the relationship between the region's prosperity and scientific productivity will be explored in greater depth, but suffice it to say, the laboratory was unable to effectively assist the industry, and turned to smaller-scale, more theoretical research. The laboratory staff, counting three to four chemists in the years of prosperity, was continually decreasing over the years, and Vèzes had to work more and more often alone with increasingly obsolete equipment.

The most lasting contribution of Vèzes and his laboratory during 1900–1920 was the publication of articles on resin chemistry in the renowned French chemical periodical *Moniteur scientifique du Docteur Quesneville*. They were later gathered and published by Vèzes in a single volume in 1922 [Vèzes, 1922]. The book constitutes the fullest and the most complete account of the advances in resin chemistry in the first two decades of the twentieth century and served as the basis for the resin chemistry compendium, *Résines et Térébenthines*, published two years later [Vèzes and Dupont, 1924].

3. The Pine Institute and the Resin Chemistry Laboratory in the Interwar Period (1922–1937)

By the middle of the 1910s, while Vèzes continued to be a respectable figure in the region, his laboratory was in decline. However, World War I brought back a combatant rhetoric similar to the one after the Franco-Prussian war. The science-industry relations became an important issue, and a new generation of practically oriented chemists tried to reshape the French research landscape by looking for synergies between laboratories and factories. Dozens of "applied sciences institutes" were established throughout the entire country, in particular in chemistry, but also in physics and related disciplines [Grelon, 1988; Pestre, 1992; Grelon, 1996]. Many of them were, first and foremost, engineering schools, but unlike their predecessors, they had full-fledged laboratories and collaborated closely with French companies. In this favourable political climate, Vèzes and his younger collaborator Georges Dupont sought to reinvigorate resin chemistry through a new research institution called the *Institut du Pin* (hereafter, the Pine Institute).

The Initial Project

In 1920, Vèzes and Dupont formulated a plea for the establishment of the Pine Institute in the local newspapers *Bois et Résineux* (roughly translatable as "Wood and Softwood", or "Wood and

Resinous Trees") and *La Forêt de Gascogne* (Forest of Gascony), which were widely read by the people working in the forest industry in Aquitaine. In their calls, the two scientists fully exploited the language widespread in France after the tragedy of the Great War:

> A diabolic genius seems to have forged for the enemy more and more terrifying and deadly weapons that have annihilated our efforts: first, the surprise of the heavy artillery, then the asphyxiating gases, then the increasingly powerful explosives ... How often in difficult hours have we regretted this scientific superiority of our enemies [Dupont and Vèzes, 1921, p. 4].

The authors argued that this scientific superiority did not end with the conclusion of the hostilities and that it could still be used against the French, as a new war — this time, an economic and industrial one — appeared to be looming on the horizon. The Pine Institute would then be a means of "economic struggle" in a new European landscape, and its research work an "act of patriotism" [Dupont, 1923, p. 35]. Vèzes and Dupont argued that the new institution was necessary for the survival of the local industry. "Considering the fact [that turpentine spirit and rosin] have a limited number of immediate uses, we notice that they had been, before the war, exported to Germany in considerable amounts ... We have no reason to leave to the Germans the benefits of this industry" [Dupont, 1921, p. 31]. The Institute's ambition was then almost political: to contribute to the French industrial and scientific independence. "We build on science, or we disappear," reiterated Dupont, explaining his belief that the origin of the German economic superiority stemmed from the strong industrial culture in German chemical laboratories. According to him, the existing resin laboratory in the premises of the Faculty of Sciences of Bordeaux was far too small, and the industry needed a new, well-equipped research unit — not "just a simple laboratory" but "a true Pine Institute" [Dupont, 1921, p. 30].

Dupont described a tentative structure that this Institute should adopt. It would comprise five sections: (1) research laboratory,

(2) technical teaching, (3) test laboratory, (4) documentation office, and (5) industrial office. The first three correspond to the roles of the resin laboratory formulated back in 1900, but the last two were new additions to the unit's profile, indicating the direction its creators envisaged for their institution.

The first section, the research laboratory, was at the heart of the project from the outset. It was meant to be divided into four units: gum harvesting and treatment, turpentine spirit and derivatives, rosin and derivatives, and wood and forest products. It is evident that the Pine Institute was to be primarily devoted to resin-related studies, and wood was more of an accessory topic for Dupont and Vèzes at the time. Another important point is that the laboratory was not meant to work exclusively on resin industry products, but the researchers were also expected to apply the methods of science to the harvesting practices of forest workers, including the study of the equipment they used.

The second section, technical teaching, was meant to educate the people willing to work in the industry, as well as those who worked there already but were looking for a more scientific approach to understand the chemical processes involved.

The third section, the analytical laboratory, would be devoted to the detection of fraudulent practices and counterfeit products, as well as to the study of raw materials. While this question is studied in detail later, it must be noted that the producers in the region faced two types of challenges. First, they had to ensure that the quality of their products remained consistent. This depended on several factors, such as weather conditions during harvest, stocking methods, and the type of gum produced. Second, the local industry had to deal with the problem of counterfeit products. Some gum producers diluted their gum with water and mixed turpentine with a cheap mineral solvent called white spirit. The control of the resinous substances, be it to detect non-fraudulent irregularities or straight-up frauds, required advanced equipment unavailable to small resin distilleries and local industry workshops. The Pine Institute was supposed to fill the gap.

The fourth section, the documentation office, was meant to gather four types of documents: scientific articles for the laboratories, industrial documentation "including the register of all patents and information helpful for our industry leaders", trade documentation, and finally, legal documentation, especially concerning fraudulent practices and information on patent breaches. In his vision, Dupont wanted to create a trade and scientific information centre focused on diffusing resin-related knowledge to the widest possible public, manufacturers, and potential clients.

The last section, the industrial office, would be "a permanent link between different sections of the laboratory and the industry. ... this office will employ one or more engineers ... familiar with the life of the industry [who would be able] to push different sections towards the most profitable outcome for the industry" [Dupont, 1921, p. 31].

Together, the five sections would form an interdisciplinary institution with an economic and scientific character. To fund such an ambitious project, Dupont wanted to create an association in which forest owners, forest workers, distillers, and all sorts of industry leaders would contribute, through membership fees, to the Institute's budget. Clearly, Dupont was aware that the establishment of such a Pine Institute would take time. As a temporary solution, its role was to be played by "the old resin laboratory" at the Faculty of Sciences of Bordeaux. In the long run, however, Dupont underlined that it was "out of the question to industrialise ... a state laboratory" [Dupont, 1921, p. 35]. He wanted it to remain focused on fundamental research — "such a laboratory has to remain, in fact, of general interest and, even if it could provide service to the regional industry, it is out of the question to reserve the benefits [of its works] to a single category of interested parties. It is then necessary that the resin industry has its own laboratory" [Dupont, 1921, p. 34].

The campaign launched by Vèzes and Dupont proved successful, and, on 24 February 1922, 24 local entrepreneurs decided to form the Pine Institute Association based on the famous French law on freedom of association from 1901.[7] It brought together a number of

[7] Departmental Archives of Gironde, Pine Institute Archives, file "Institut du Pin de faculté". "Statut de l'Association Institut du Pin", no date, probably 1935.

influential political figures, entrepreneurs, and the wealthiest forest owners [Dupont, 1923]. The Association rapidly became the most important stakeholder of the entire local industry. Its statute announced in article 1 that the main functions of the newly created entity was to "stimulate, study, centralise, coordinate, develop and encourage everything that can, directly or indirectly, be of interest for [the cultivation of] resinous plants and [for the development of] their products". What kind of tools did the Association have to achieve this goal? — "Organisation of conferences, classes, and expositions, establishment of scholarships, funding technical publications and specialised laboratories".[8]

In practice, however, the Association struggled to form any coherent strategy on how to develop the local industry due to incessant internal conflicts between its members, though the directors of the pine resin laboratory could develop it almost unbothered. They grasped the opportunity, stepped in, and took the lead of both the local science and the local resin industry.

The Institute's Founders

Vèzes' vision for the Pine Institute was developed by his successor as the head of the university resin laboratory, Dupont (Figure 4). The transition between the two directors was gradual. Due to health problems, Vèzes slowly gave way to his assistant, but he remained involved in the activities of the Pine Institute until the end of his days. Much like Vèzes, Dupont had studied at the very prestigious Parisian *École normale supérieure* (ENS) under the direction of the future Nobel Prize laureate, Paul Sabatier.

At his own initiative, Dupont contacted Vèzes to help with his work concerning the resin industry immediately after World War I. In 1921, in the midst of their campaign for the establishment of the Pine Institute, he became an assistant professor at the Faculty of Sciences of Bordeaux. In 1930, he was offered the position of the

[8] Departmental Archives of Gironde, Pine Institute Archives, file "Institut du Pin de faculté". "Statuts de l'Association Institut du Pin", no date, probably 1935.

Figure 4: Georges Dupont (1884–1958), the first technical director of the Pine Institute and one of the leading French chemists working on terpenes.

Source: Family photo from the private archives of Guy Rapin, CC BY-SA 4.0 (https://commons.wikimedia.org/wiki/File:Georges_Dupont.jpg)

dean of the same faculty, where he remained until 1933, when he left Bordeaux and returned to Paris to become the director of the laboratory of chemistry at the ENS. In 1948, Dupont became the school's general director, one of the most coveted positions for a scientist in France [Charle, 1989, p. 21].

Would this brilliant career have been possible had the Pine Institute project failed? What is clear is that, thanks to the Pine Institute, Dupont became a respected figure in the scientific and industrial community. His fame began in 1921, when he designed a distillation column for pine resin, making it possible to manufacture products considerably purer than before [Dupont, 1926, p. 55]. He was also actively involved in research on resin properties and applications.

Between 1924 and 1933, Dupont authored or co-authored at least a hundred articles on resin and wood chemistry. He also continued to publish articles in the field every now and then after his departure to Paris. The numbers alone are impressive, but Dupont's scientific approach was also original. He combined physical chemistry, a hot area of research at the time, with real-life problems encountered in industrial chemistry. He employed the newest discoveries, such as Raman spectroscopy, to help the resin industry control its products [Lestel, 2008, p. 179–186].

Contrary to Vèzes, who began his career in inorganic chemistry and gradually moved towards resin chemistry, resin chemistry remained only a fraction of Dupont‘s activities — he also studied many more theoretical subjects. Today, he remains most well-known for his contributions to terpene chemistry, which he undertook for the first time as early as the 1920s, precisely during his stay in Bordeaux [Lestel, 2008, p. 179–186].

Despite being the true founder of the Pine Institute, Dupont did not hold the position of general director of the Pine Institute's laboratory. Instead, he was appointed the "technical director". After the resignation of Vèzes in 1925, the position of director of the resin laboratory was filled by a certain Laurent Richard, who was also the director of the previously mentioned School of Chemistry Applied to Industry and Agriculture, a major chemical engineering school in Bordeaux originally created by Ulysse Gayon. Nothing suggests that Richard was involved in the management of the resin laboratory, and he never published on resin chemistry. He was most likely responsible for administrative questions and for the integration of the unit into the academic landscape of the region, while Dupont managed the Institute on a day-to-day basis.

In 1933, when Dupont departed for Paris, Richard retired as well. Vèzes remained the honorific director (this was his official title), but he was unable, due to his age and health, to manage the laboratory. Dupont, before his departure, designated a successor: "I will, undoubtedly, be replaced by my young and knowledgeable colleague, M. Truchet, an ENS [*École Normale Supérieure*] graduate ... He will be helped by M. Brus, one of my former students, whose Ph.D. dissertation concerned turpentine spirit." [N. N., 1933b] However, in the following year (1934), it was Brus who became the technical director, with no one taking up the position of director, strictly speaking (Truchet became an honorific director though).

Born in 1899 in Castres, a little village in the department of Tarn, Georges Brus defended his thesis on pinene and nopinene in 1929 at the Faculty of Sciences of Toulouse. Pinene and nopinene are two terpenes of turpentine spirit extracted from *Pinus maritima* resins. Dupont called Brus one of his former students, and effectively, they published a few works together. Their first article on pinonic acid appeared in April 1923 in *Annales de chimie* (Annals of Chemistry) [N. N., 1933b]. Despite the geographical distance (he was working on his Ph.D. in Toulouse), Brus was a close co-worker of the Institute in the 1920s, and he published the entirety of his dissertation in the pages of the Pine Institute's bulletin [Brus, 1929–1930]. After having spent years studying turpentine, there is no doubt that he was a competent specialist in the field, but he had little previous administrative experience and was "only" a provincial university graduate, unlike Truchet, who finished the prestigious Parisian ENS.

Why was it Brus and not Truchet who became the new director after the departure of Dupont? The archives remain silent, but a short anonymous note from the period explained that "the director, professor of the Faculty of Sciences, was not designated by the Faculty".[9] Perhaps the industry leaders from the Pine Institute Association favoured Brus, someone who had been involved in the industry for more than a decade instead of an outsider from Paris, no matter how talented this person was. Another possible explanation relates to the fact that in 1934, Truchet became the director of the School of Chemistry Applied to Industry and Agriculture in Bordeaux. It is possible that he decided to concentrate exclusively on this task, leaving the Institute to Brus. In any case, Brus took the position and held it for no less than three decades. Like Dupont before him, he became Dean of the Bordeaux Faculty of Sciences in 1944. After 1944, even though he kept the title of director of the Pine Institute, his role

[9] Departmental Archives of Gironde, Pine Institute Archives, file "Institut du Pin de Faculté". "La situation actuelle de l'Institut du Pin", no date, probably around 1935/36, p. 4.

gradually decreased, and it was the Institute's technical director, Pierre Legendre, employed in 1941, who became responsible for its everyday management, very much like Dupont in the 1920s. In the late 1960s, after Legendre's retirement, Brus again became directly involved in the Institute's activities, but he himself retired in 1969.

Brus is one of the most important protagonists in the story of the Pine Institute, and it is crucial to note the breadth of his activities. He was a scientist and lecturer at the Faculty of Sciences of Bordeaux, but he also served as the primary consultant for the entire resin industry as well as for the community of chemists involved in resin chemistry in Europe. Throughout his tenure, he received hundreds of inquiries from various individuals and companies regularly seeking his advice on different aspects of the Institute, such as its profile (some believed it to be a beauty and health school due to the historical use of turpentine as a remedy[10]), the products of local companies (mistaking the Pine Institute for a chamber of commerce[11]), or the properties of resin products (for example, whether resin-based camphor could help against a rabbit infestation[12]). All of this was part of the everyday work of the Institute's director.

Staff, Students, and the International Fame of the Pine Institute's Laboratory

Three scientists and talented administrators, Vèzes, Dupont and Brus, deeply marked the history of chemistry in the south-west of France. But the history of the Pine Institute was, of course, not just limited to the history of its directors. The resin laboratory of the Pine Institute, in the interwar period, was a place of lively encounters between the

[10] Departmental Archives of Gironde, Pine Institute Archives, file "Correspondance 1937 G. B., T–U". Letter from M. Tamisier, 29 February 1937.

[11] For example: Departmental Archives of Gironde, Pine Institute Archives, file "Correspondance 1937 G. B., E". Letter to Manufacture Liégeoise de Caoutchouc, 27 May 1937.

[12] Departmental Archives of Gironde, Pine Institute Archives, file "Correspondance 1951–1952, P". Letter from Produits & Matériels Agricoles, 1 December 1952.

research staff, students, and foreign guests. For the first group, it is necessary to underline that collaborators of the resin laboratory had varying statuses. Many chemists from the Faculty of Sciences collaborated with the Institute without being paid directly from its budget, especially during 1924–1937. In fact, the resin laboratory did not even remunerate some of its permanent employees; for example, in 1923, the *Institut des recherches agronomiques* (Institute of Agronomical Research (IRA)) gave an allowance to the Institute to pay for two chemists: Marcelle Barraud and Roger Uzac [Brus and Menaut, 1938, p. 85]. The IRA was replaced by the national *Service de la répression des fraudes* (Fraud Repression Service) in 1934. Other institutions also occasionally funded scholarships in the Institute, such as the National Office of Liquid Fuels [N. N., 1927c]. Other members of the laboratory included assistant professors of chemistry, who split time between teaching at the faculty and working in the laboratory itself.

In 1937, at least fourteen people were somehow involved in the activities of the resin laboratory, not counting students. How was their work divided? Between 1924 and 1938, the Institute had five sections: (1) Harvesting and treatment of gum resin, (2) Turpentine spirit, terpenes, and derivatives, (3) Rosins and derivatives, (4) Wood and paper (with two subsections: wood distillation and paper), and (5) Analyses. This division closely mirrored the one proposed by Dupont in the early 1920s. Each of them had a section head, but some chemists managed two sections at the same time.

One of numerous scientists, Marcelle Barraud, who worked in the premises of the laboratory, deserves a little more attention. She worked at the resin laboratory from 1922 to 1941 and was the head of two sections for the entire period: Turpentine and Analysis. In 1928, she defended her Ph.D. dissertation, which systematised knowledge on turpentine spirit and its properties. Throughout her career at the Institute, she published 27 papers in the Institute's bulletin, of which she was also the main editor for many years. In 1935, she was awarded the title of *chef de travaux* (chief chemist) at the Pine Institute, created specifically in recognition of her significant contributions. Considering that turpentine was the most expensive and important product of the industry, and that Barraud worked closely with local producers on the quality control of all their products, she held a privileged position at the intersection of the local scientific and industrial landscapes. In 1933, at a banquet organised for Dupont just before his departure to Paris, he reassured the entrepreneurs in his farewell speech that "Mlle Barraud stays here. She knows inside and out, as you know, everything that concerns resinous trees and products" [N. N., 1933].

In parallel to her strictly scientific studies, she was also involved in the debates on the economic future of the Landes region. She advocated for a narrower trade definition of turpentine to protect local manufacturers [Barraud, 1930] and was interested in the history of the economic development of the forest [Barraud, 1930b]. She was also directly responsible, alongside Brus, for responding to correspondence sent to the Pine Institute, especially concerning trade and standard-related questions. Finally, she also regularly participated in the local scientific and economic events.[13]

And yet, despite having played such an eminent role in the local science and industry, Barraud was never paid by the Institute. Her scholarship was first funded by the IRA and then by the Fraud Repression Service, but her employment was irregular and precarious. In 1933, during the banquet, Dupont called for creating a permanent position for Barraud, considering her current status unacceptable, but nothing seemed to have changed after he left. In fact, in 1941, following a restructuring of university services, Barraud was detached from the Pine Institute and placed in the Centre of Agronomic Research of Grande Ferrade (working on wines), of which she became a director around 1947. The archives remain silent about the circumstances of her departure from the Institute.

Two opposing views concerning Barraud can be formulated. On the one hand, the Institute might be considered a progressive place

[13] Departmental Archives of Gironde, Pine Institute Archives, file "Correspondance 1938 M. B., A". Letter from Association Française pour l'Avancement des Sciences, 19 October 1938; Departmental Archives of Gironde, Pine Institute Archives, file "Correspondance 1938 M. B., C". Letter from Comité Départemental de la revue Bois et Résineux to M. Barraud, 9 February 1938.

where a woman could become a section head and conduct a successful pioneering career in an environment dominated by men. On the other hand, her inability to secure a permanent position and her exclusion from consideration as a potential director, despite having the requisite qualifications, highlights the difficulties faced by female scientists in French provincial universities during this period.

Directors and laboratory leaders such as Barraud aside, the laboratory of resin chemistry at the Faculty of Sciences of Bordeaux was also a temporary home to researchers from many places, a phenomenon particularly evident when analysing the student population. The Pine Institute activity reports, while incomplete, indicate that the most prolific year was 1927 when no less than thirteen Ph.D. students of six nationalities were working within the Institute's walls, with five of them completing a new engineering Ph.D. (unlike a traditional Ph.D. in chemistry) that was introduced in France only in 1923 [Courtebras, 2008, p. 100] (Figure 5).

Figure 5 illustrates the remarkable degree of internationalisation of the Institute, which gained worldwide recognition and attracted students interested in the exploitation of forest resources not only from neighbouring countries facing similar challenges (in terms of climate and forestry practices), such as Portugal and Spain, but from countries as far as Canada and China.

In fact, researchers from East Asia contacted the Institute every now and then for information on resinous products; for example, in 1937, after an official request from the Embassy of Japan, the Institute hosted two researchers, a wood chemist from the Imperial Ministry of Agriculture and Forests and another scholar from the Imperial Faculty of Osaka.[14] Another region where the Pine Institute achieved recognition was Eastern Europe; for example, in 1927, four Serbian Ph.D. students conducted their research projects within the premises of the Institute. In 1937, the Soviet Society for Intellectual

[14] Departmental Archives of Gironde, Pine Institute Archives, file "Correspondance 1937 G.B., A". Letter to the Japanese Embassy, 23 June 1937; Departmental Archives of Gironde, Pine Institute Archives, file "Correspondance 1938 G.B., H". Letter from N. Hirao, 28 May 1938.

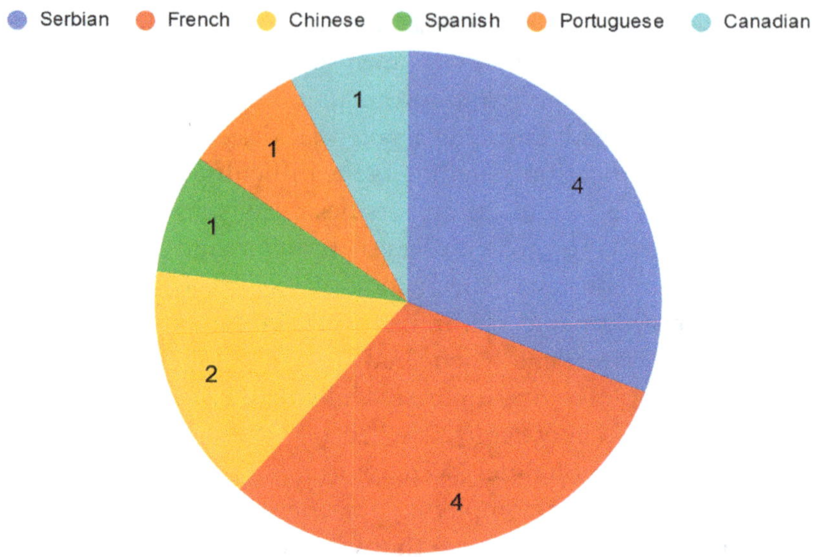

Figure 5: The Pine Institute's Ph.D. students by nationality (1927).[15]

Exchanges contacted the Institute to publish comments on the Soviet scholarship on resins in the Institute's bulletin.[16] In 1938, one Polish engineer from Lviv applied no less than three times for a temporary position in the Institute in order to discover its techniques and laboratory practices.[17] All these examples reiterate that despite a relatively modest number of permanent employees, the Institute was a lively place full of students, interns, and foreign visitors, and that its scientific contributions were recognised worldwide.

[15] Established from data in [N. N., 1928, p. 49].

[16] Departmental Archives of Gironde, Pine Institute Archives, file "Correspondance 1938 G.B., K". Letter from Kimiska Fabrika "Mersrags", 14 May 1938.

[17] Departmental Archives of Gironde, Pine Institute Archives, file "Correspondance 1938 M. B., S". Letter from T. Sawicki, 24 September 1938; Departmental Archives of Gironde, Pine Institute Archives, file "Correspondance 1938 G. B., S". Letter from Związek Inżynierów Chemików Rzeczypospolitej Polskiej, 6 October 1938; Departmental Archives of Gironde, Pine Institute Archives, file "Correspondance 1938 M. B., S". Letter from T. Sawicki to M. Barraud, 8 November 1938.

Teaching and Knowledge Popularisation

Research might have been the primary mission of the Institute, but its scientists were also actively involved in communication on resin chemistry to a wider public. As already explained, the history of pine resin chemistry started in the Landes, with Vèzes giving classes to the local industry leaders. In the interwar period, the Pine Institute continued this tradition by offering open lectures at the university for a diverse audience. Topics covered in these lectures included turpentine, rosin, paints and varnishes, and wood, paper, and even forest fuels.

In parallel with university teaching, the staff at the Pine Institute were actively involved in various outreach initiatives; for example, in 1941, Brus participated in "Earth Day" celebrations and prepared an exposition on the economic importance of the forest for people living in the rural areas in the Landes.[18] Another project involving the Pine Institute was the *École pratique de l'industrie du pin* (Practical School of the Pine Industry), whose mission was to "form educated labourers". These labourers would have spent some time in the school and attended lectures at the Faculty of Sciences. They were to acquire the necessary skills for professions, such as a forest worker specialised in resin harvesting (*résinier*) or a distiller in a resin distillation workshop. The school was meant to include a small experimentation service co-directed by the Pine Institute. In fact, the members of the Pine Institute Association were intended to be part of the school's council. Alas, the school was not established despite support from a special commission established by the French Ministry of Public Education. Most likely, the project became too expensive after 1927 when a major crisis hit the entire resin industry [Dhommée, 1927, p. 226].

Commercial Side of the Pine Institute

Research and teaching aside, the Pine Institute provided several fee-based services to local entrepreneurs. In a certain way, it acted as if

[18]Departmental Archives of Gironde, Pine Institute Archives, file "Correspondance 1941, G". Letter from Brus to the town hall of Saint-Magne, 5 August 1941.

it was a private company, or what is called today a shared service centre, where entrepreneurs can either order chemical analyses of resin-based products (and many more, in fact) or simply buy an appropriate equipment to conduct these quality tests themselves. How did this activity work in practice? Already in 1924, the resin laboratory established a loyalty programme of sorts for those contributing financially to its budget. Those who paid a fixed fee had guaranteed free analyses of resinous products and a subscription to the Institute's bulletin, as well as favoured treatment when it came to patent licensing. The most important advantage was, however, priority treatment for inquiries on resinous products and their properties. In 1937 alone, the Pine Institute was consulted no less than 1,000 times on issues ranging from sources of rosin colouration to the average prices of turpentine on local markets.

A considerable number of these requests concerned the analysis of resinous products. The vast majority of small workshops, distilleries, and local companies in the region lacked the laboratory equipment and expertise necessary for even the most rudimentary quality tests. These analyses were essential for revealing potential defects in raw materials, such as resin diluted with water, and in distillation products, such as turpentine with high impurity content. Not only were these tests critical for providing quality products, they helped to protect consumers against fraudulent practices. In fact, the laboratory was frequently consulted by local entrepreneurs and the companies with they traded with, for example, paint and varnish factories that were dissatisfied with the turpentine they had acquired. The Pine Institute often acted as an arbiter, providing expertise that was later used in courtrooms. Table 2 provides an illustration of the typical services offered by the Institute.

Although the exact nature of these analyses and their methodology have evolved since 1924, the main controlled characteristics were the same in 1924 as in the 1930s and the 1960s.[19] However,

[19] Departmental Archives of Gironde, Pine Institute Archives, file "Lettres Techniques 1964–1965, E". Letter from S.A. Eurocolor, 29 June 1964.

Table 2: Typical analyses conducted by the Pine Institute laboratory in 1924 [N. N., 1924d]

Test type and price	Material	Characteristics
40 francs — Complete analysis	Turpentine spirit	Physical properties (density, refractive index, optical rotation), acidity, oil resin and rosin adulteration, other adulterants
	Resin	Composition (amount of terpenes, acids, water, and impurities)
	Dry products (rosin)	Amount of turpentine, acidity, impurities
20 francs — analysis	Turpentine	Major fraud detection, rosin, and resin oil adulteration
	Resin	Water and solid matter dosage
	Dry products (rosin)	Amount of residual turpentine
	Resin oil	Viscosity and boiling point
12 francs — analysis	All	Identification of one property

a notable change occurred in the range of products analysed over time. Already, in the 1930s, the analytical services of the Institute extended to wood, fuel, and other materials not always directly related to the resin industry, such as coal. In a way, the Institute was looking for new markets to conquer and strengthen its place in the industrial panorama of Bordeaux. The fees charged for analyses represented 20 per cent of the Institute's budget in 1938, and although this proportion slightly decreased after World War II, the analytical laboratory remained the centrepiece of the Institute until the 1960s.

Who were the clients? For example, Joseph Farbos, one of the most important producers in the region, ordered at least 89 tests between 1937 and 1972, mostly on turpentine and resin oil. Another important customer was *Compagnie des Produits Resineux* (Company of Resinous Products), which, over the same period, ordered at least 77 analyses. However, as already mentioned, the Pine Institute used its talents to help other branches of industry as well; *e.g.*, *Forges et Chantiers de la Gironde* (Forges and Dockyards of Gironde) ordered at least a hundred analyses concerning steel, fuel oil, and lime. The *Association des Propriétaires d'appareils à vapeur du Sud-Ouest* (Association of Steam Apparatus Owners of South-West) solicited the Institute regularly for studies on their coal and water.[20] In this context, the most faithful client of the Institute should be mentioned: the Fraud Repression Service, a state agency with vast regulatory competence, especially in agriculture, ensuring the quality of the products on the market. This long-lasting collaboration had started already back in 1908 when a Ministry of Agriculture decree provided that "The Resin Laboratory was officially designated to conduct analyses of samples of resinous products taken on the entire French territory" [Brus and Menaut, 1938, p. 86]. From this date, the Fraud Repression Services regularly sent samples to the Bordeaux laboratory. Most of them concerned adulteration of the most expensive product of resin distillation: turpentine, often diluted with cheap white spirit.

The archives do not provide the exact count of analyses conducted every year in the interwar period by the Institute, but for the years for which there are more exact numbers (1954–1963), the analyses for the Fraud Repression Service represented around 25 per cent of the total, which varied from around 600 (1954) to around 1,600 (1961).[21] These numbers were certainly lower in the 1920s and the 1930s, although the Institute's correspondence indicates that it was often overwhelmed with requests and some companies waited months for results. To address the issue, the Institute started selling basic equipment, making it possible to conduct simple analyses in the distillation workshops themselves.

---

[20] Departmental Archives of Gironde, Pine Institute Archives, files "Correspondance" between 1937 and 1972.

[21] Estimates based on U.C.R. yearly reports in *Bois et Résineux*, 1955–1963.

Figure 6: Device to measure the water in a resin sample.
Source: Courtesy of Ecomusée de Marquèze-PNRLG.

Equipment Sold by the Pine Institute

The most widespread and simplest fraud committed by forest workers consisted of adding water to gum to increase the overall volume of the harvest. The presence of a certain amount of water was unavoidable due to primitive harvesting methods, but the law defined brackets that an unadulterated resin should adhere to. The Pine Institute sold a toolbox allowing anyone to conduct this kind of test and evaluate whether the water content did not exceed legal requirements [N. N., 1933c] (Figure 6).

Another instrument invented by the Pine Institute was used to measure the acidity of turpentine spirit (Figure 7). Resin distillation involves separating volatile terpenes (turpentine) from solid rosin made of resinic acids (similar to fatty acids). However, poorly distilled turpentine can contain a proportion of acids, which are

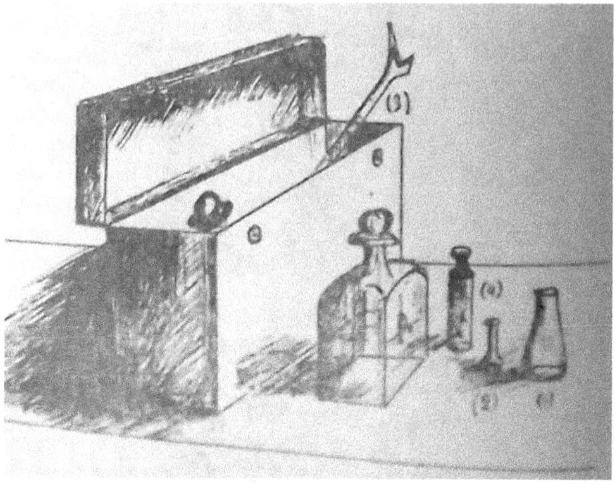

Figure 7: Tool for dosing acidity of turpentine.
Source: *Bulletin de l'Institut du Pin*, 40, 1933, p. 78.

considered adulterants, and the turpentine's acidity can make it unfit for many applications, especially as a solvent.

The last tool sold by the Pine Institute was for evaluating the quality of the solid products of distillation, namely rosins. In French, bright rosins are called "*colophanes*", and the darker ones are called "*brais*". Together, they were collectively labelled as dry products of resin distillation. In fact, for a long time, the price of the solid resin depended entirely on its colouration — the brighter, the better, especially for paints and varnishes and in soapmaking. Darker colours meant the presence of a significant proportion of polluting substances. In order to quickly evaluate the shade of rosin, a reference model was established. It was made of glass slides, the colours of which corresponded to the ones of rosin with different degrees of adulteration. The slides alone were not enough for an exact measurement though. A special tool, a colophoscope (Figure 8), was manufactured by a technician from Bordeaux and sold via the Institute.[22]

[22]Departmental Archives of Gironde, Pine Institute Archives, file "Lettres Techniques 1964–1965, V". Letter from Volkart Brothers, 10 September 1956.

Figure 8: Colophoscope, the tool for measuring the colour of rosin samples, and the official French glass rosin colouration set.

Source: Departmental Archives of Gironde, Pine Institute Archives (photos).

4. Towards the Pine Institute of the Faculty of Sciences of Bordeaux (1937)

The previous sections have highlighted the remarkable accomplishments, which garnered both national and international renown, of the Pine Institute. As a result, it became a magnet for students and resin chemistry scholars worldwide, drawn to Bordeaux by its prestigious reputation. Yet, amidst this tale of success, a certain ambiguity arises. Indeed, it had been the laboratory of resin chemistry at the Faculty of Sciences that was behind most of these achievements. Contrary to Dupont's initial wishes, no new independent industrial laboratory was created. However, this situation was unclear to external stakeholders who believed that the resin laboratory was simply "the" Pine Institute.

In fact, there were four separate and independent entities that represented the Pine Institute to the outside world and whose members overlapped, giving the impression that they all belonged to the same organisation (Table 3 and Figure 9). These entities were: (1) the Pine Institute Association, which consisted of local entrepreneurs, politicians and scientists, and which collected funding for research, (2) the Society of Studies and Applications for the Progress of the Resin Industry, which managed intellectual property and sold equipment developed within the resin laboratory, (3) the Pine Institute Bulletin, a semi-autonomous publication funded by the Ministry of Agriculture and managed by the laboratory, and (4) the university resin laboratory itself that actually conducted research.

Table 3: Chronology of the Pine Institute institutional environment (1921–1940)

<table>
<tr><td>'21</td><td>'22</td><td>'23</td><td>'24</td><td>'25</td><td>'26</td><td>'27</td><td>'28</td><td>'29</td><td>'30</td><td>'31</td><td>'32</td><td>'33</td><td>'34</td><td>'35</td><td>'36</td><td>'37</td><td>'38</td><td>'39</td><td>'40</td></tr>
<tr><td colspan="17">Resin Laboratory (1901–1937)</td><td colspan="3">"New" Pine Institute</td></tr>
<tr><td></td><td colspan="19">Pine Institute Association (1922–middle 1940s, unimportant after 1937)</td></tr>
<tr><td></td><td></td><td></td><td colspan="16">Pine Institute Bulletin (1924–1939)</td><td></td></tr>
<tr><td></td><td colspan="16">Society for the Progress of the Resin Industry (1922–1937)</td><td></td><td></td><td></td></tr>
</table>

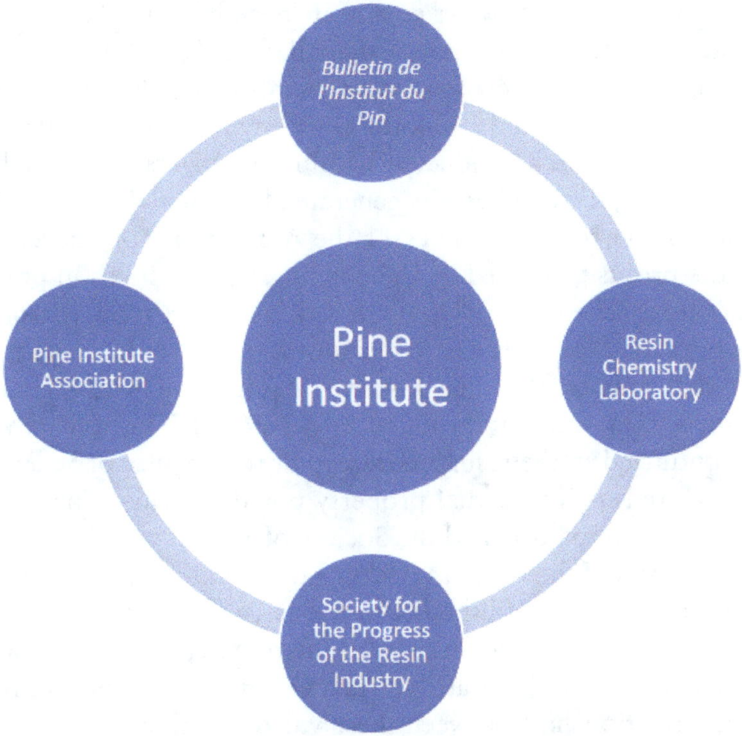

Figure 9: Rendition of the core Pine Institute components between 1924 and 1937.

The Association was "the" Pine Institute established in 1922. It was intended to be the steering institution of the new organisation with its own independent industry-funded laboratory but failed to achieve its ambition. Notably, in 1922, Dupont explained that the Institute was to become self-sufficient thanks to the intellectual property licensed to local companies [Dupont, 1922, p. 2]. The problem was that the legal form adopted by the Pine Institute Association did not allow for any kind of commercial activity. Four months after its creation, on 19 June 1922, a trading company was established: the *Société d'études et d'applications pour le progrès de l'industrie résinière* (Society of Studies and Applications for the Progress of the Resin Industry). Its mission was "to implement the discoveries of the Pine Institute [as imagined in the 1920s], take patents, negotiate their licensing, and pay back 80 per cent of its benefits to the laboratory of the Pine Institute".[23] The initial capital was 100,000 francs, half brought by Pine Institute Association members and half from selling shares. Between 1922 and 1934, the company filed around 30 patents (based on the research conducted within the university resin laboratory). However, the real benefits generated by its activities are hard to evaluate, as the budget outlines of the resin laboratory never mentioned the profits generated by the patents [Vèzes, 1934]. In practice, the laboratory relied, above all, on grants and direct funding from the members of the Pine Institute Association.

To recapitulate, the Pine Institute Association was a civil law association that did not produce any patents. The research was solely conducted in the resin laboratory at the Faculty of Sciences of Bordeaux, and its intellectual property portfolio was managed by a small private company called the Society of Studies and Applications for the Progress of the Resin Industry. Additionally, the Pine Institute Bulletin, funded and published with the assistance of the Ministry of Agriculture, presented the findings of the research carried out within the university resin laboratory, further adding to the confusion about the relationships between these various entities.

From an organisational and hierarchical standpoint, the relationship between these entities was messy, and their very members often failed to grasp their boundaries and respective competence. An anonymous internal report from the mid-1930s retraced the evolution of the problem:

> [T]he university laboratory continued then to develop itself ... not under the name of "The laboratory of chemistry applied to the resin industry", but under the name "The Pine Institute". ... The Pine Institute we dreamed of was never created. It is "the resin laboratory" of the Faculty of Sciences that became "the Pine Institute". But the current situation of the laboratory ... is completely irregular from the administrative point of view.[24]

[23] Departmental Archives of Gironde, Pine Institute Archives, file "Institut du Pin de Faculté". "La situation actuelle de l'Institut du Pin", not dated, probably 1935/36, p. 2.

[24] Departmental Archives of Gironde, Pine Institute Archives, file "Institut du Pin de Faculté". "La situation actuelle de l'Institut du Pin", not dated, probably 1935/36, p. 2.

As previously explained, Dupont opposed industrialising the university resin laboratory, which was intended to remain in the public interest. However, as no new industrial laboratory was established specifically for the Pine Institute, the resin laboratory at the Bordeaux Faculty of Science continued to fill this gap. A temporary solution thus became a permanent one. Brus was aware that the situation of the laboratory was precarious since the members of the Pine Institute Association believed they had the right to set up the laboratory's research strategy (as they considered the laboratory to be merely an extension of their organisation). A short anonymous notice from 1935 explicitly pointed out the nature of the misunderstandings between the Association and the laboratory: the leaders of the Association "do not realise that even if most of the subventions come through their hands, ... this is only because people know that they would go to a faculty laboratory ... They do not understand that the fame of the Pine Institute [is due to the works of the resin laboratory]".[25]

At one point, in the middle of the 1930s, there were deliberations to rename the Pine Institute Association as "Patronage Committee of the Pine Institute" or even "Friends of the Pine Institute", so that the laboratory could simply be called the Pine Institute. However, Brus was sceptical and predicted that the Association members would oppose this solution.[26] An internal report from 1935 stated that regularising the relations between the laboratory and the Association should be considered a priority because the Association was often considered an unwelcome intruder by the Bordeaux Faculty of Sciences, for which the resin laboratory was simply one of many of its chemistry laboratories with no special status. According to the report, the Faculty should recognise and accept the symbolic authority of the Association members over the laboratory to avoid tensions. In order to satisfy both the Association leaders and the faculty's directorship, a new organisational formula had to be developed.

---

[25] Departmental Archives of Gironde, Pine Institute Archives, file "L'Institut du Pin de Faculté". "Note sur l'Institut du pin", 1935.

[26] Departmental Archives of Gironde, Pine Institute Archives, file "Institut du Pin de Faculté". "La situation actuelle de l'Institut du Pin", not dated, probably 1935/36, p. 5.

The inspiration came to Brus from Lille as he decided to model the new Pine Institute on the *Institut de la Houille* (Coal Institute).

The Coal Institute was established in 1931 and was funded by local coal mines, the French National Office for Liquid Fuels, and the Lille Faculty of Sciences.[27] It was hosted in the university building and comprised multiple laboratories working on geology, chemistry, paleobotany, and petrography. Its administrative council included all the previously mentioned stakeholders, alongside major industry leaders. Every year, the council established the main lines of research that were thereafter explored by the Coal Institute's laboratories. The legal form chosen for the Coal Institute was the "Faculty Institute", introduced by the decree on universities from 31 July 1920. Article 3 of the decree explained that the universities were allowed to establish "Institutes destined for scientific research or for [the study of] practical applications". Article 5 added that these institutes could be governed by administrative councils with non-academic members.

It remains unclear why the decree in question had not been used earlier in 1922 in Bordeaux to establish the Pine Institute. However, in the middle of the 1930s, the Coal Institute was all over the internal memos of the resin laboratory, and it was directly cited in the debates concerning the reform of the Pine Institute. The documentation on the experiences from Lille was presented to the Pine Institute Association members. The internal notice of the laboratory explained that "[t]he form of the Coal Institute pleases [the members] ... Those gentlemen were seduced by the idea that the administrative council of the Coal Institute establishes every year a programme responding to the needs of the interested parties and presents it to the director".[28] The lengthy negotiations led to a convention between the Pine Institute Association and the Faculty of Sciences of Bordeaux signed on 21 May 1937 that established a new institution: the Pine Institute of the Faculty of Science, which was meant to replace the old resin laboratory. The convention outlined three principal missions for the

[27] Departmental Archives of Gironde, Pine Institute Archives, file "L'Institut du Pin de Faculté", "L'Institut de la Houille de la Faculté des Sciences de Lille", p. 1.

[28] Departmental Archives of Gironde, Pine Institute Archives, file "L'Institut du Pin de Faculté". Note sur l'Institut du pin, 1935, p. 2.

institution (which were slightly less ambitious than the one in 1922): (1) conduct scientific research aiming for practical applications of all products derived from pines, (2) study all questions interesting for the development and the protection of the industries exploiting these products, and (3) train engineers and chemists for these industries.[29]

The stipulations of the convention regularised the relations between the Association and the Faculty without radically altering them, and it may have appeared as a mere rebranding exercise: the university resin laboratory simply changed its name to the Pine Institute. In practice, however, the change in power relations was far-reaching. The newly created Institute gained considerable autonomy from both the Faculty and the Association. First, the Institute's director could limit the influence of the Faculty's dean by invoking the new special status of the semi-autonomous "faculty institute". Second, the laboratory formally acquired the right to use the name "Pine Institute", meaning that the potential stakeholders, clients and people interested in funding its activities had no need to go through the Association anymore, making the latter redundant.

On 27 October 1937, the ministerial decree approved the convention, and the new organisation was formally recognised by the French administration. Its main organ was an administrative council of fifteen members, who were the representatives of the Faculty, public administration, local businesses, including four members of the Pine Institute Association.[30] Interestingly, in the Pine Institute archives, there are no minutes of the meetings of the council, nor any

---

[29] Departmental Archives of Gironde, Pine Institute Archives, file "L'Institut du Pin de Faculté". "La convention entre l'Association de l'Institut du Pin et la Faculté des Sciences de Bordeaux", 21 May 1937.

[30] The members were the dean of the Faculty of Sciences of Bordeaux, the Pine Institute's director, two professors or lecturers chosen by the Faculty, the president of the Pine Institute Association alongside three other delegates, the general director of the Waters and Forests Administration, the director of the Forestry School in Nancy, the local conservator of waters and forests, the general director of the Fraud Repression Service, the general inspector of the French technical schools, the president of the Federation of the Unions of Resin and Wood-based Product Producers of Landes, and finally the president of the Departmental Committee of Wood and Resins of Gironde.

testimonies concerning its activities. This lack of evidence leads to questions about whether the council actually functioned or not. Additionally, the Pine Institute Association disappeared during the war, leaving the Institute's special status owed to a convention between two parties, one of which eventually vanished. Consequently, the responsibility for representing the Institute to the outside world, determining its research strategy, and making decisions on budgets and employment rested solely with the Pine Institute's director. This position was held by the main author of the convention, Brus, for more than 30 years, between 1934 and 1969. In a sense, the reform meant to solve the irregular situation of the laboratory in the 1920s and the 1930s, created a new ambiguity, which was maintained in the following decades. Despite the significant changes in the Institute's activities and profile after the disappearance of the resin industry in the 1970s, its administrative foundations remained unchanged. No reforms or reorganisations in the second half of the twentieth century affected the Institute's administrative structure. In other words, from a formal standpoint, the Pine Institute, as established in 1937, remained in existence until 2010.

5. The Pine Institute Bulletin

The previous sections explored the tangled development of the structure that became the Pine Institute. This section proposes to reread the history of the Pine Institute between 1924 and 1939 through its major publication organ, *Bulletin de l'Institut du Pin* (Pine Institute Bulletin). The Bulletin was a monthly (and later bi-monthly) review published between 1924 and 1939. It was created to consolidate the articles previously published in various journals and produced by the laboratory's scientists. The aim was to make the Bulletin the most comprehensive publication on resin chemistry and the resin industry in France and to be the primary publication outlet for the laboratory's scientists. From 1924 to 1928, the editor of the Bulletin was the biochemist Eugène Aubel from the Laboratory of Physiological Chemistry, which was not a part of the Pine Institute. From 1929 up to 1939, it was Barraud, the leader of the turpentine section of the

Pine Institute, who was the chief editor of what became the largest publication in the world on resin chemistry.

According to the editorial in its first issue from April 1924, the Bulletin's mission was to create:

> [A] collection of technical notes destined not only to keep the subscribers informed about [the Institute's] works but also to give them, in condensed form, summaries of all the publications in the world that can deliver some knowledge on resinous wood and resinous products and show their new uses [Dupont, 1924c].

This editorial explained that the totality of knowledge on resin chemistry produced by the resin laboratory so far was assembled in three books: *Résines et Térébenthines* (Resins and Turpentines) (1924) by Vèzes and Dupont [Vèzes and Dupont, 1924], *Distillation du bois* (Wood Distillation) (1924) and *Essences de térébenthine* (Turpentine Spirits) (1926) by Dupont [Dupont, 1924b; Dupont, 1926]. The Bulletin intended to go beyond these works and expand the frontiers of the discipline as established by these three books.

In practice, however, the Bulletin was not limited to resin chemistry or pine wood chemistry but provided a platform for discussions on economic and even political questions (such as fraud legislation) of interest to the resin industry. Figure 10 shows, for example, an article written by Barraud on the economic history of the industry.

Every issue of the Bulletin was divided into two parts. The first included original research articles, and the second, called "documentation", was made of summaries of articles published in other journals, mostly from abroad. A bibliometric analysis of the Bulletin shows numerous interesting developments throughout its existence.[31] First, it is worth noting that the Bulletin published, on average, 34 original articles a year, including longer articles published as parts over several issues. When it comes to the proportion of original articles, they

[31] The entire scientometric analysis of the Bulletin was conducted in collaboration with Alexandre Saffre, whom I acknowledge for his contribution.

N° 12. (2e Série) Paraissant le 15 de chaque mois. 15 Décembre 1930.

BULLETIN DE L'INSTITUT DU PIN

Sous le contrôle de l'Institut des Recherches agronomiques et rattaché à la Faculté des Sciences de Bordeaux

A I 40

HISTORIQUE DU Développement Economique des Landes

Causerie faite à l'A.R.C.A. le 14 décembre 1930 par Mlle Marcelle Barraud

Il est un fait acquis, c'est que le pin maritime existe dans le sud-ouest de la France depuis la plus haute antiquité. Le géographe romain, Pomponius Mela, et le médecin grec Dioscoride, au commencement de l'ère chrétienne, parlent de la poix et de la résine que les Romains tiraient des Gaules. En effet, avant l'ère chrétienne, une florissante cité gauloise, Boïos, existait non loin de La Teste-de-Buch. Les habitants de cette cité, les Boïens, très laborieux, avaient déjà trouvé le moyen d'arrêter l'envahissement par les sables de la mer, par des semis de pin maritime. Ils faisaient un fructueux commerce de produits résineux avec l'Espagne et l'Armorique, ce qui leur valut, de la part de saint Paulin l'épithète de « Piceos », noirs de poix. Au début du Ve siècle, les Barbares ravagèrent la riche Aquitaine, saccagèrent Boïos, incendièrent les forêts, dont la dépouille fut rapidement enfouie sous la mer et les dunes de sable. Par suite de la formation des dunes, les ruisseaux n'eurent plus libre cours vers la mer, les étangs peu à peu se formèrent le long du littoral, recouvrant bientôt la grande voie romaine qui aurait existé entre Bayonne et Boïos. Dans cette plaine marécageuse, hérissée d'îlots de pins maritimes, a vécu, jusqu'au XIXe siècle, une population misérable.

La forêt de La Teste, vestige des richesses de Boïos, d'une part, la forêt du Marensin, d'autre part, demeurèrent longtemps les deux seuls centres importants du commerce des résineux. Un marché de matières résineuses avait lieu deux fois par semaine à La Teste; il en est fait mention dans les rôles gascons de 1382 et 1383; on retrouve également un règlement applicable à la vente et à l'entrée des résines à Bordeaux, en 1415, des chartes relatives à des chargements de pièces de résine, de barils de térébenthine destinés à Orléans, Rouen, Bilbao, Londres, Anvers, au cours du XVIe siècle. Au début du XVIIe siècle, Bayonne fait aussi un commerce fructueux de matières résineuses, alimenté par l'important marché de Dax.

On vit uniquement sur la forêt existante; aucune tentative n'est faite, semble-t-il, par les gouvernants, pour accroître la forêt, et lutter contre l'envahissement du pays par les sables. Montaigne rapporte que les habitants disent que « depuis quelque temps la mer se poulse si fort vers eux qu'ils ont perdu quatre lieues de terre... ». Des efforts individuels sont tentés, cependant pour vaincre le sable. Nous retrouvons dans les archives de Bayonne, par exemple, que le Conseil de la ville, réuni le 17 mai 1639, désigne M. de Labat, jurat, pour se porter sur les sables d'Anglet et « faire esmonder les pins que la ville y a fait en divers ans ensemencer pour arrester les sables... ». A la même époque, les habitants de Labenne ayant partagé leurs communaux, déclarent dans l'acte, que « chacun y ferait et complan-

— 1 —

Figure 10: First page of Issue 12 of *Bulletin de l'Institut du Pin* published on 15 December 1930.

represented, on average, 76 per cent of the pages of the Bulletin, the rest being devoted to reprints and summaries.

All articles published in the Bulletin (original or not) were divided into seven categories: A. Generalities, B. Resin harvest and treatment, C. Turpentine spirit, terpenes, and derivatives, D. Rosin and derivatives, E. Wood chemical derivatives, F. Cellulose, and G. Diverse documents. The first and last category aside, the remaining categories largely corresponded to big research themes of the Institute. If grouped into resins (categories B, C, D), wood (E, F), and others (A, G), the first one corresponds, on average, to 55–60 per cent of the content, the second one to around 20–25 per cent, and the last one to 20 per cent. However, this varied from year to year (Figure 11).

The exact evolution of every category is due to a number of factors. The occasional spikes often corresponded to singular events, such as a publication of a dissertation or conference proceedings, but some broader trends, for example, the primacy of resins over wood in certain periods, are clearly identifiable. What matters is that the main theme of the Bulletin was pine resin chemistry (and resin chemistry-related challenges), which was studied in three categories, B, C and D, corresponding to resin harvesting, turpentine and rosin,

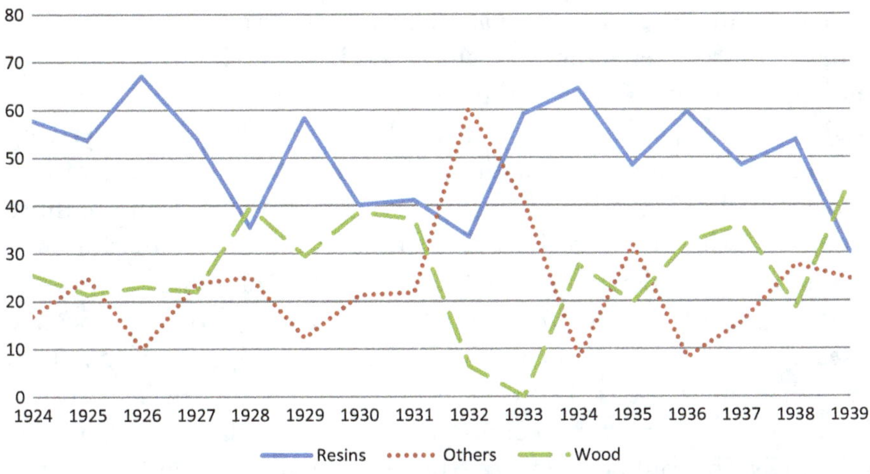

Figure 11: Percentage of *Bulletin de l'Institut du Pin's* volume in a given category.

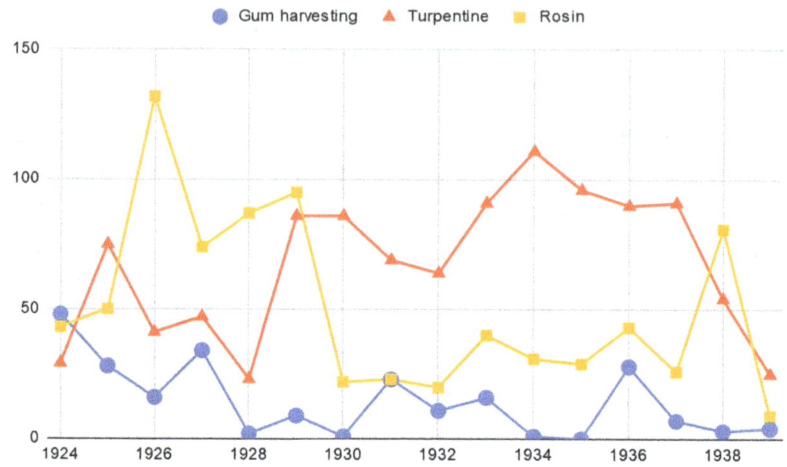

Figure 12: Pages of *Bulletin de l'Institut du Pin* in resin-related categories.

respectively. Figure 12 shows the evolution of their volume (in terms of the number of pages).

According to Figure 12, resin-harvesting methods were considerably less important than other topics and became almost marginal after the first years of the Bulletin's existence. Nevertheless, the topic never really disappeared, showing the will of the Institute to contribute in a domain that was as close to forestry as to chemistry. What is more interesting: turpentine replaced rosin as the topic of choice in the second half of the 1920s. It appears that the economic crisis that ravaged the region over the period was the underlying reason for this shift. When the prices of resin and its derivatives plunged in 1927, the Pine Institute researchers focused on the more expensive distillation product: turpentine. Another explanation may be related to the departure and arrival of Dupont and Brus in 1934 as the head of the Institute and the influence of their respective research interests. Dupont worked on a wide range of eclectic topics, while Brus focused almost exclusively on turpentine and on writing articles about the economic situation of the Landes. Figure 13 shows the publications of the two directors divided by the Bulletin's categories.

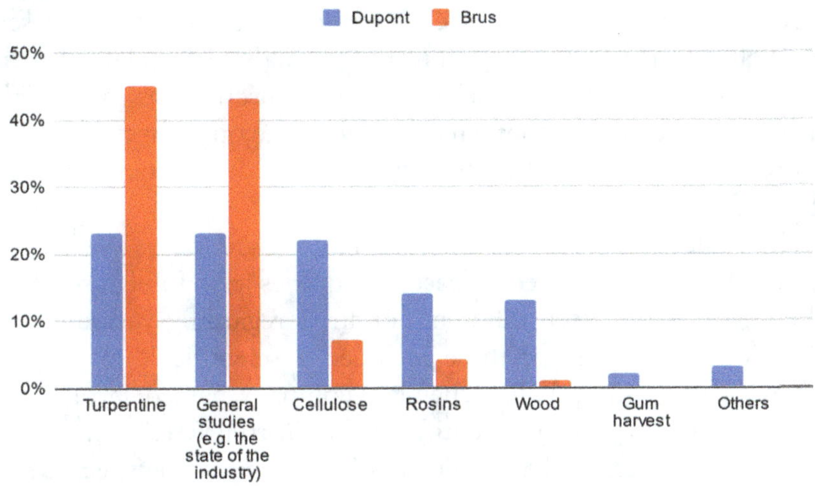

Figure 13: Proportion of Dupont's and Brus's publications on a given topic published in *Bulletin de l'Institut du Pin* (1924–1933).

The two directors also differed in terms of productivity. Brus published considerably less in the pages of the Bulletin and often preferred to publish in general reviews in chemistry such as *Bulletin de la Société chimique de France* (Bulletin of the French Chemical Society). In contrast, Dupont published frequently in the Pine Institute Bulletin and continued to occasionally publish there even after he left the Institute in order to stay in touch with the local scientific community.

Who were the Bulletin's readers? Local industry leaders and French specialists in resin chemistry first, without a doubt. But among its regular subscribers, there were also chemists and engineers working in many other industries. Among them was the chemical giant Rhône-Poulenc[32] as well as smaller entities, such as the General Union of French Brewery Syndicates[33] and the famous perfumer

[32]Departmental Archives of Gironde, Pine Institute Archives, file "Correspondance 1938 G. B., Q-R". Letter from Rhône-Poulenc, 1 March 1938.

[33]Departmental Archives of Gironde, Pine Institute Archives, file "Correspondance 1937 G. B., Q-R". Letter from Union Générale des Syndicats de la Brasserie Française, 20 March 1937.

Roure-Bertrand Fils from Grasse.[34] Very often, the Bulletin was sent free of charge in exchange for internal journals and bulletins of the companies in question. These regular subscriptions aside, the Institute was regularly contacted by the companies that wanted to acquire specific issues of the Bulletin that contained articles relevant to their activity.

The Bulletin had gained international recognition as well. From 1937 to 1940, Brus received at least 25 inquiries from foreign scientists and companies in fifteen different countries regarding various issues and articles. Examples of these included a scientific library in the Netherlands[35], a small Polish chemical company named Jarot,[36] and the British Department of Scientific and Industrial Research.[37] More often, individual researchers contacted Brus about some specific articles, *e.g.*, the Nobel laureate Leopold Ružička from ETH Zurich[38] and the initiator of biochemistry in India, Professor Biresh Chandra Guha from the University of Calcutta.[39] In these exchanges, Brus often acted as a tradesman trying to sell his stock; for example, when a certain Fred Schulz from Prague wanted to buy a single issue of the Bulletin to get an overview of resin products' uses, Brus convinced him to acquire the entire collection (more than 120 issues) to get a full picture.[40]

If it was so successful, why did the Bulletin disappear? Its last issue was printed in February 1939. Two subsequent issues

[34] Departmental Archives of Gironde, Pine Institute Archives, file "Correspondance 1937 G. B., Q-R". Letter from Roure-Bertrand Fils, 15 December 1937.

[35] Departmental Archives of Gironde, Pine Institute Archives, file "Correspondance 1937 G. B., V-W". Letter from Wetenschappelijke Boekhandel, 16 April 1937.

[36] Departmental Archives of Gironde, Pine Institute Archives, file "Correspondance 1937 G. B., T-U". Letter from Jarot, 16 January 1937.

[37] Departmental Archives of Gironde, Pine Institute Archives, file "Correspondance, 1937 M. B., F". Letter from British Department of Scientific and Industrial Research, 18 January 1938.

[38] Departmental Archives of Gironde, Pine Institute Archives, file "Correspondance, 1937 G. B., Q-R". Letter from Leopold Ružička, 4 November 1937.

[39] Departmental Archives of Gironde, Pine Institute Archives, file "Correspondance 1937 G. B., G". Letter from Prof. Guha, 14 June 1937.

[40] Departmental Archives of Gironde, Pine Institute Archives, "Correspondance 1937 M. B., S". Letter from F. Schulz, 21 October 1937.

(March–April and May–June) were printed but not distributed "during the war because of their content", according to Brus, although it remains unclear what he referred to.[41] Possibly, it might have been due to the nationality of the authors — some were Germans; for example, see [Schnurrenberger, 1939]. Financial difficulties might also have accounted for putting the Bulletin to rest. In 1937, in a letter to *Société chimique de France* (French Chemical Society), Brus confessed that the Bulletin faced financial troubles and asked the Society for a subvention.[42] The editors of the Bulletin also struggled with respecting delivery deadlines. The issues were notoriously printed and distributed long after their official date of publication (this happened, for example, in 1939), leading to complaints from the subscribers. The local printing house was blamed, but it appears that there were also underlying financial troubles.

This does not mean that the concept of the Bulletin was abandoned and that Brus somehow sabotaged it. In a letter dated 26 November 1941, he indicated that the Bulletin should "resume in a few months".[43] In fact, throughout the war, interest in the publication did not fade: Brus was regularly contacted by clients asking whether there were new issues. In June 1942, Brus explained in a letter to the French publisher Hachette that the publication would be renewed after the end of hostilities.[44]

And yet, it did not happen. In the subsequent years, Brus was repeatedly contacted by different stakeholders asking whether new issues were already out, and Brus answered every single time that the Bulletin was about to be renewed, but no specific date was given,

---

[41] Departmental Archives of Gironde, Pine Institute Archives, file "Correspondance 1939, D". Letter of G. Brus to Decremps et Conquet, 19 December 1939.

[42] Departmental Archives of Gironde, Pine Institute Archives, file "Correspondance 1937 G.B., D". Letter from G. Brus to Delaby, 22 May 1937.

[43] Departmental Archives of Gironde, Pine Institute Archives, file "Correspondance 1941, L". Letter from G. Brus to Libraire de Chimie Industrielle, 26 November 1941.

[44] Departmental Archives of Gironde, Pine Institute Archives, file "Correspondance 1942, H". Letter of Département Etranger Hachette, 17 June 1942.

often complaining about publications costs.[45] In June 1956, the technical director Pierre Legendre put an end to all speculation and finally confirmed that the Bulletin would not be renewed.[46]

After the Bulletin ceased to be published, however, the Institute continued to receive orders for past issues up until the 1960s.[47] Decades after the last issue, the Bulletin was thus still coveted around the world. The fact is that *Bulletin de l'Institut du Pin* was a unique reference in the field of resin chemistry on a global scale. No other publication on resin chemistry achieved the same status, leaving a significant gap after World War II. In a certain way, the Bulletin was the pillar of the entire discipline of resin chemistry, and without it, the discipline itself struggled to affirm its specificity. This will be further discussed in the following chapters.

Conclusions: Pine Institute as a Hybrid and Modular Entity

When examining the history of the Institute, it becomes apparent that its establishment was not marked by a single founding moment. The establishment of the resin laboratory in 1900 was a step towards integrating science and industry in the region and paved the way for

[45] Departmental Archives of Gironde, Pine Institute Archives, file "Correspondance 1946, M". Letter from La Maison Rustique, 16 December 1946; Departmental Archives of Gironde, Pine Institute Archives, file "Correspondance 1949, E". Letter from Filipe Jorge Mendes Frazao, 24 June 1949; Departmental Archives of Gironde, Pine Institute Archives, file "Correspondance 1950, B". Letter from J. Ballereaud, 3 October 1950.

[46] Departmental Archives of Gironde, Pine Institute Archives, file "Correspondance 1956, A". Letter A. Letter from Académie Yougoslave de Sciences et des Beaux-Arts, 7 June 1956.

[47] Departmental Archives of Gironde, Pine Institute Archives, file "Lettres Techniques 1964–1965, R". Letter from Raffinerie Tirlemointoise S.A, 30 July 1965; Departmental Archives of Gironde, Pine Institute Archives, file "Correspondance 1968, K". Letter from K. Kokkinos, 29 April 1968; Departmental Archives of Gironde, Pine Institute Archives, file "Correspondance 1968", Letter S. Letter from Swets & Seitlinger, 23 December 1968.

further collaborations, but the laboratory's role shrank over the years. The year 1922 was a new beginning, rekindling the interest in resin chemistry and solidifying the discipline in the region. The Pine Institute that emerged in the 1920s was an organic, modular, and constantly evolving institution. The resin laboratory at the Bordeaux Faculty of Sciences fulfilled its research function. The Pine Institute Association provided the permanent link between scientists and the local industry. The trade company (Society of Studies and Applications for the Progress of the Resin Industry) diffused the practical applications stemming from resin-related research to potential clients, and the Pine Institute Bulletin played the role of communication office. The Pine Institute, as understood in the interwar period, was, in fact, defined by the relations between these four entities. It was not planned for the Institute to have this unintuitive form; it was an emergent organisation that came out of the interactions between these four bodies. Despite this lack of clear boundaries, the Pine Institute was still perceived as a single entity by outsiders.

And yet, its "messiness" was problematic for the members of the constituent bodies themselves. Unclear boundaries led to the restructuring and formation of a new Pine Institute at the Faculty of Sciences of Bordeaux in 1937. It is important to underline that the rich modular character of this instance of the Pine Institute started waning precisely after 1937. The trade society disappeared the same year, and two years later, in 1939, the Pine Institute Bulletin was discontinued. The Pine Institute Association lost its position of importance after the war when a new and much larger entity, *Union Corporative des Résineux* (Corporative Union of Resinous Products), became the Institute's major interlocutor in terms of funding, although it had a very different, more general, character.

It is important at this point to reconsider the broader scientific context in which the Institute could emerge. According to the historian of science Harry Paul, central actors such as Gayon, Vèzes and Duhem formed, alongside other Bordeaux scientists, a generation of pioneers, that instead of reproducing academic elites, "trained laboratory personnel for faculties and chemists for governmental agencies in agriculture, customs, war, colonies, and railways, as well as for industries in agriculture, gas, and metallurgy" [Paul, 2003, p. 199]. The Pine Institute, driven by the ambition of the following generation of scientists, including Dupont and Brus, was a consequence of this particular ideological setting in which the practical application of science and the collaboration with the industry were set as a model to follow.

This local trajectory was fully in line with similar trajectories all over the world. The Pine Institute was what can be called an industrial research institute, a type of "hybrid institution" that emerged in the late nineteenth and the early twentieth centuries, functioning at the frontier between science, industry, and public administration. These institutes produced scientific expertise not in service of a single company but of entire branches of industry, simultaneously equipping the state with tools to regulate various activities through standardisation. The category of industrial research institutes is indeed far from uniform. It includes major national entities, such as the *Physikalisch-Technische Reichsanstalt* (Imperial Physical Technical Institute), established in Berlin in 1887, and the British National Physics Laboratory in Teddington (1900), but also smaller ones, more specialised or closer to the industries, such as the American (1929) and the French (1919) optics institutes. One of the closest, thematically, to the Pine Institute was the *Pappersmassekontoret* (the Swedish Association for Pulp Research) from 1917 [Kaiserfeld, 2013, p. 183].

This is to say that understanding the process of creating an organisation such as the Pine Institute involves a number of concomitant factors. On the one hand, the Institute's establishment is part of broader international tendencies, but it cannot be solely explained by them. On the other hand, while the Institute's history is local, it is not separable from the general zeitgeist, often through direct inspiration from other institutions, such as the Coal Institute of Lille or the Viticulture Station of Bordeaux. It is crucial to incorporate this persisting multilayered feature of scientific institutions into every historical analysis to avoid oversimplified explanatory models that focus only on either the global or the local dimension.

Chapter 2

CRAFTING A DISCIPLINE: RESIN CHEMISTRY IN AQUITAINE IN THE FIRST HALF OF THE TWENTIETH CENTURY

> "*La chimie est à la fois science et art. Science, parce qu'elle procède par des lois rigoureuses dans l'analyse comme en synthèse. Art, par le côté imaginatif de sa conception constructive.*"
> Desalbres, L. (1946). La vie financière de l'Union Corporative des Résineux, *Bois et Résineux*, 30 November 1946, p. 1.

The particularity of resin chemistry as an autonomous discipline lies in the fact that it was embedded into a very narrow economic and institutional setting. Although chemists from around the world have examined resin from different perspectives, the creation of a single disciplinary matrix was mainly achieved through the publications in the Pine Institute Bulletin and the activities of the Pine Institute. Consequently, the development of resin chemistry was inextricably linked to the difficulties faced by the forest industry in Aquitaine. Local firms provided the laboratory with practical challenges and ways of thinking about resin that moulded the field for decades. This chapter delves deeply into this intricate connection through a number of narrow technical case studies illustrating key difficulties encountered by scholars interested in pine resin chemistry. These case studies are classified into three primary research categories: turpentine spirit, rosin, and raw gum resin. In each of these areas, scientific challenges of either theoretical or practical importance can be identified. The selected examples do not reflect the Institute's complete range of activities, but they present the core of the Institute's contributions in the field.

This chapter comprises five sections, of which the first two lay the groundwork for subsequent discussions. The initial section provides a general introduction to the history of resin chemistry, whereas the second discusses the scientific and industrial roots of resin chemistry in Aquitaine prior to the establishment of Maurice Vèzes' laboratory. The last three sections explore the three major research trajectories: gum extraction, turpentine and rosin, as well as the problems studied by the Pine Institute and its contributions to the local industry within each area.

1. What is Resin?

Answering this question is not a straightforward matter. Alexander Tschirch, a Swiss-German pharmacist renowned for his pioneering work on resin, explained in the early twentieth century that defining this heterogeneous substance posed significant challenges: "The notion of resin is neither chemical nor medical, but physical, taken from the everyday experience, from a series of particular properties that made resin convenient for the manufacture of varnish." Resin's definition appears to be linked to its applications. For Tschirch, resinous "opposes a certain resistance to the touch, ... shows a strong cohesion of the substance, a sort of internal friction, ... [it is] neither fatty, nor oily, nor elastic". According to him, the resin's main characteristics were "insolubility in water; solubility in alcohol or ether; softening by heating ...; evaporation of the solution gives a transparent lacquer; a relative resistance to reagents, to putrefaction, as well as to digestive juices of the body" [Tschirch, 1926, p. 508]. This definition illustrates a certain ambiguity present in all discussions on resins — the criteria listed by Tschirch are satisfied not only by a number of natural substances but also by synthetic ones.

Synthetic resins find wide usage in different branches of the industry, and terms like "polymer", "resin", and "plastic" are often used interchangeably. The origins of synthetic resins can be traced back to Adolf von Bayer, who, in 1872, observed the poly-condensation of phenol with formaldehyde. In the following decades, many more chemists, such as Alexander Parkes, Louis Blumer, and especially Leo Baekeland, with his famous bakelite, successfully introduced synthetic resins to consumers [Mercelis, 2020]. This book does not explore the rich history of synthetic resins, but it must be noted that these new substances immediately challenged the monopoly of natural resins in the lacquer and varnish industry [Wilks, 2001, p. 1129].[48] The Pine Institute studied synthetic resins exclusively in the context of using pine resins as raw feedstock for their production.

Even if we exclude synthetic resins from our study, the definition of various plant resins remains difficult, as there is no simple intuitive way to divide them into mutually exclusive categories. René Lombard, a chemist from Strasbourg who collaborated closely with the Pine Institute, identified four frameworks for classifying resins in his seminal work *Produits résineux, gemmes, colophanes & dérivés* (Resinous products, gums, rosins and derivatives): pragmatic (traditional), chemical, botanical, and rational (Table 4). These frameworks diverge from each other and present distinct elements, but they also occasionally overlap [Lombard, 1946, p. 3]. Lombard's work highlights the epistemological challenges involved in defining resins.

These different frameworks are rooted in disciplinary ways of thinking. Botanists tend to focus on the origin of the substance, whereas chemists are more concerned with its composition, and

[48] Note that natural resins should not be exclusively associated with plant resins. Traditionally, the term "natural resins" also encompassed shellacs, which are secretions produced by *Kerria lacca*, an insect native to India and Thailand. Historically, shellac found use in the production of phonographic discs. However, with the advent of synthetic resins, shellac lost its prominence in this application and is now predominantly utilized in woodworking projects [Hicks, 1961].

Table 4: Four frameworks for classifying resins according to René Lombard (1946)

Pragmatic (traditional)	Oleoresins (different turpentines, Canada balsam, elemi), Resins proper (amber, copal, dammar, colophony, accroides, mastic, sandarac, Dragon's blood), Balms (from Mecca, Peru and Tolú, benzoin, styrax, storax, sweetgum, copaiba), Gums (from Senegal, arabic, nostras gums, tragacanth), Tannifer gums, Gum-resins (gamboge, Asa foetida, galbanum, sagapenum, myrrh, frankincense, opoponax, aloe), Latex
Chemical	Ester resins, Resene resins, Resin acid resins, Resinol resins, Aliphatic resins, Glucoresins, Chromoresins, Enzyme resins, Lactoresins, Gums, Tannifer gums, Gum-resins, Latex
Botanic	Gymnosperm phanerogams (copals), Turpentines (pines, firs, larches), Amber, Bourgogne pitch, Angiosperm phanerogams (Dragon's blood, aloe, acroid resins, rubber, gamboge), Dipterocarpus (camphor, dammar, Gurjun balm), Terebinthaceae (Chio turpentine, mastic, myrrh, Japanese lacquer, frankincense, elemi), Legumes (gums, balms, copals, *etc.*).
Rational (by Lombard)	Conifer secretions (*e.g.*, turpentines), Copals and Copahu oleoresin, Dipterocarpus resins, Terebinthaceae resins, Papilionaceous resins, Gum-resins, Saxifrages resins, Styracaceae resins, Other resins, Animal resins, Gums, Latex.

industrialists or pharmacists prioritize the practical properties. The challenge arises when the knowledge associated with these different substances is generated independently within these different disciplinary matrices, leading to the development of competing terminologies for the same substances. Not only that, some practitioners borrowed terms from other disciplines, often confusing incompatible vocabularies and reproducing previously existing imprecise traditional terminologies. In practice, any researcher interested in the topic of resins had, and still has, to navigate in a sea of definitions to uncover their meaning through comparative studies. It should be noted that Table 4 was translated from French, and some of the names do not correspond exactly to common English names. To get a full picture, one would have to understand the evolution of uses of every single one of these names in different languages, as well as of their circulation between them.

The pine resin itself can serve as an example of this confusion (illustrated in Table 4). According to the pragmatic classification, pine resin is simultaneously present in two separate categories: turpentine (understood as the volatile part of distillation or a very liquid resin) is considered an "oleoresin", while colophony (rosin) is a "resin proper". At the same time, in the botanical framework, all pine resins are simply called "turpentines", and in the chemical one, "resin acid resins". Lombard's rational framework categorises pine resins as turpentines but only as a subset of the larger group of "conifer secretions". Another example of the confusing terminology is copal resins, which are sometimes considered resins and legumes at other times.

The nomenclatural chaos can be an important challenge; for example, in medical professions when it comes to the identification of allergic reactions to resins. One of the medical articles exploring the problem of imprecise definitions explained that "the literature on resins, although relatively abundant, is not very precise ... we must note the extreme confusion existing with regard to [the] copals and damars, confusion which, in a way, reflects the complexity of the world of resins" [Sell and Foussereau, 1989, p. 229]. A problem with the definition of damars was also noted in the literature on varnishes and lacquers [Feller, 1964].

For the most part, the debates about different types of resins and how to classify them have been extensively explored in classic textbooks of resin chemistry. Alexander Tschirch's *Die Harze und die Harzbehälter* (Resins and Resin-producing Structures) [Tschirch, 1900] laid down the foundations for resin studies. This book was followed by the aforementioned famous *Résines et Térébenthines* by Vèzes and Dupont [1924], and then Lombard's *Produits résineux, gemmes, colophanes & dérivés* [Lombard, 1946]. In the English-speaking world, the first overarching compendium on resins from the chemical standpoint was Frank Norman Howes' *Vegetable Gums and Resins* (1949) [Howes, 1949]. All these books were written in the first half of the twentieth century, when resin chemistry flourished as an independent discipline. But when its institutional foundations eroded in the 1950s and the 1960s, grand works trying to explain and organise knowledge in the field were discontinued. Of course, the works on diverse resins continued in different disciplines, but the overarching view was lost. This gap was partially filled by a new eclectic approach of ethnobotany. In 2003, the famous botanist Jean Langenheim published her compendium *Plant Resins: Chemistry, Evolution, Ecology and Ethnobotany*, which remains the Magnum Opus on this complex topic [Langenheim, 2003]. Langenheim proposed a much narrower definition of resin than Lombard. In her book, resin is:

> [D]efined operationally as a primarily lipid-soluble mixture of volatile and nonvolatile terpenoid and/or phenolic secondary compounds that are (1) usually secreted in specialised structures located either internally or on the surface of the plant and (2) of potential significance in ecological interactions [Langenheim, 2003, p. 24].

Her definition combines botanical and chemical considerations. This narrow approach makes it possible to distinguish resins from sister substances such as gums and latex (Table 5).

It is necessary to point out that in many older classifications, including the ones of Lombard and Tschirch, latex and gums were subtypes of resin and not independent substances. This is not surprising considering that both are extracted in a similar manner to resins and share similar physical characteristics. It should be remembered that the word "gum" (*gemme* in French) was and still is widely used in the pine resin industry to describe resin. In English, the term "gum spirits of turpentine" has a strict definition in trade standards.

These terminological debates and the extent to which they were entrenched in local contexts will be explored later in the book. Retrospectively, however, what matters is that many scientists working on resin chemistry in the first half of the twentieth century concentrated specifically on pine resin due to its economic

Table 5: Characteristics of resins and their sister substances [Langenheim, 2003, p. 45]

	Primary components	Solubility	Secretory tissue
Resins	Terpenoids, phenolic compounds	Lipid soluble	Canals, pockets, cavities, trichomes, epidermal cells
Gums	Polysaccharides	Water soluble	Cavities
Mucilages	Polysaccharides	Water soluble	Idioblasts, epidermal cells, trichomes, ducts, cavities
Oils (fats)	Fatty acids and glycerol	Lipid soluble	None
Waxes	Fatty acids esterified with long-chain alcohols	Lipid soluble	Unspecialised epidermal cells
Latex	Complex mixture, may include terpenoids, phenolic compounds, proteins, carbohydrates, *etc.*	Lipid soluble	Laticifers (elongated secretory cells)

importance. In a certain way, pine resin was a resin par excellence, serving as a benchmark for other definitions. These pine resins were often described as oleoresins that, according to Langenheim, "are comparatively fluid terpenoid resins. They have a higher ratio of volatile to non-volatile terpenes than some other resins, such as balsams and copals" [Langenheim, 2003, p. 306]. In the works of chemists from the first half of the twentieth century, the terms "pine resin", "pine gums", and "oleoresin" were usually used interchangeably.

It is worth noting that pine resins are "terpenoid" substances, which means that their molecular structures contain carbon backbones made up of isoprene units. Isoprene is a volatile, unsaturated organic compound produced by plants and animals. It contains five carbons, and therefore, the number of carbon atoms in any terpenoid is a multiple of five. Terpenic chains can be constructed from isoprene molecules, and their molecular formulas follow the general formula $(C_5H_8)_n$ for $n > 1$. For instance, the molecular formula

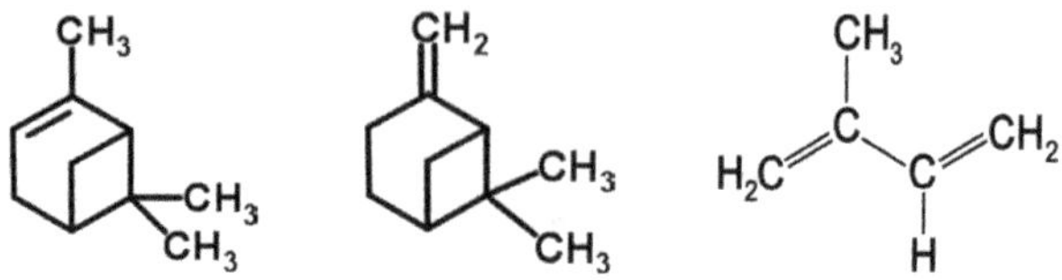

Figure 14: Simplified structural formula of α-pinene (left), β-pinene (middle), and isoprene (right).

of monoterpenes is $C_{10}H_{16}$, of sesquiterpenes, $C_{15}H_{24}$, of diterpenes, $C_{20}H_{32}$, and so on. The two most important terpenes in the volatile part of resins extracted from maritime pines in the Landes (*Pinus pinaster*) are two monoterpenes: α-pinene and its isomer β-pinene (Figure 14).

The maritime pines, which are native to the Mediterranean region, are widespread in the Landes Forest, but α-pinene and β-pinene are present in various proportions in other species, such as *Pinus sylvestris* and the American longleaf pine, *Pinus palustris*. It is important to note that different dominant terpenes may be present in other species of pine.[49]

The relationship between the conifer resins and terpenoids is constitutive to the both of them. The German chemist Friedrich August Kekulé coined the name "terpene" in 1866 to describe the compounds he isolated from the turpentine spirit of pine resin [Croteau, 1998, p. 322]. Terpenes owe their name to turpentine. At the same time, Langenheim defines turpentine resin today as a substance that contains terpenoid compounds. In the nineteenth century, it was the phenomenological reality of tangible substances in everyday life that dictated the naming conventions of chemical compounds; today, it is the molecular composition that differentiates resins from

[49] Gironde Deparmental Archives, Gironde Departmental Archives, Pine Institute Archives, box "Normalisation ISO", file "Norme France". G. Brus, P. Legendre, "Rapport sur la normalisation de l'essence de térébenthine", 20 May 1949, p. 5; for a more detailed overview of the contemporary consensus, see [Kopaczyk *et al.*, 2020].

gums, latex, and other similar substances (which used to be often seen as almost identical).

2. Industrial and Scientific History of Resins in Aquitaine

Pine resins and their derivatives have been used since antiquity to glue arrowheads, manufacture torches and candles, coat the inside of the cork stoppers in amphorae, and make barrels waterproof [Aufan and Thierry, 1990, p. 6]. This last property made them essential for the conservation of liquids (giving wines a particular taste) and also in naval construction. Tar and pitch from pine wood carbonisation were used to caulk boat hulls (to make them impermeable) and to impregnate cordage and sails (to make them water-resistant and insusceptible to putrefaction). They were also used to seal leaks while on the high seas.

Before the invention of steam ships, resin products, called "naval stores" in English,[50] played a tremendous role in international trade. Tar barrels were a typical element of harbour landscapes until the nineteenth century [Earley, 2004, p. 85]. The rulers of France were fully aware of the importance of resinous products as well. It was in the seventeenth century when this branch of industry experienced considerable growth in France after Jean-Baptiste Colbert, the famous Minister of State, decided to make Bordeaux a military and commercial centre that would establish French maritime power [Aufan and Thierry, 1990, p. 50]. Under his leadership, great shipyards of Bordeaux were constructed for which both wood and naval stores were necessary. These two raw materials came from small

[50] Note that "naval stores" is a uniquely English term. The French never had an equivalent expression. In order to maintain consistency throughout the book, I have opted for the term "resinous products", although the English-speaking literature often employed the term "naval stores" to describe various pine products long after their naval applications had declined.

pine woods (*pignadas* in the local dialect) scattered through the malarial swamps south of Bordeaux. The locals used huge kilns (*dolia*) to manufacture tars in a wood carbonisation process normally used for the manufacture of charcoal. The question that preoccupied the minds of local producers was how to obtain a purer tar than the one imported from Scandinavia, known for its excellent quality. In 1663, two specialists from Sweden were invited to Bordeaux to teach local artisans more efficient manufacturing methods [Sargos, 1997, p. 178].

At the margin of this booming industry, another activity slowly gained popularity: pine tapping (*gemmage* in French). It involves cutting into the live outer layer of a tree to create a vertical wound, referred to as a "face", which excretes viscous, shining and colourless droplets of resin. The gum was separated from water and adulterants by heating it to create turpentine (*térébenthine* in French). As time passed, focus shifted towards obtaining purer forms of turpentine: turpentine spirit (*essence de térébenthine* in French). The English and French terminologies differ somewhat, as discussed more extensively in Chapter 6. However, it is crucial to note that during the economic boom in Aquitaine, initiated by Colbert at the end of the seventeenth century, pine resins first piqued the interest of French scholars. Nicolas Lémery, the author of the first French manual of chemistry, identified two uses for resins: pharmaceuticals and varnishes. Hence, expertise on pine resin arose in two distinct communities: pharmacists and chemists in service of the industry [Lémery, 1675, p. 485]. Lémery's insights were further developed by scholars such as Louis de Jaucourt, who wrote an article on turpentine in Diderot and d'Alembert's *Encyclopédie* in 1751 [Jaucourt, 1751, p. 147–150]. But more significantly, Lémery's insights were taken up by pioneers in the field of varnish chemistry, including an Italian scholar Filippo Buonanni in his seminal work *Trattato sopra la vernice detta communemente Cinese* (Treatise on the varnish commonly called Chinese) [1731, p. 41], and a Swiss chemist Pierre François Tingry [1803, p. 45], whose *Traité théorique et pratique sur l'art de faire et d'appliquer les vernis* (Practical and theoretical treaty on the art of applying varnishes) published in 1803 was a reference work for decades.

Despite this growing body of knowledge on resins, some of the great chemists who undertook extensive laboratory studies on resins for the first time, among others (Jöns) Jacob Berzelius, the founder of modern Swedish chemistry, and Otto Unverdorben, a German chemist who discovered aniline, quickly faced a number of difficulties. Unlike their forerunners, who thought about resins and turpentines mostly through botanical categories, these scientists tried to identify their main chemical compounds. By distilling resin samples, Unverdorben obtained three materials he called "turpentine oil", "pinic acid", and "sylvic acid". According to him, the proportion of these substances could vary greatly depending on the species (be it pines, firs, larches, and other conifers); for example, the volume of turpentine oil varied from 5 per cent to 25 per cent of the distilled resin's weight [Berzelius, 1831, p. 440, 466].

Berzelius read Unverdorben's works on resins but encountered inconsistencies with the results of his own experiments. He made a remarkable point:

> I cannot explain this discordance [in the results] otherwise than by admitting that we operated on the resins that were not at all identical, despite having the same name. It is possible if we think about differences that the same resins may have depending on their age, the time of year, and the climate of the country where they were extracted, and depending on the plant species from which they were obtained [Berzelius, 1831, p. 440].

Since the time of Berzelius, it has been understood that the composition of resinous products is influenced by a multitude of variables. The confusion surrounding the terminology used during that period can be exemplified by Berzelius's own classification, where he grouped fir resin and *Pinus sylvestris* resin together and larch resin and maritime pine resin (*Pinus pinaster*) in another category. In other words, according to Berzelius, the resin from maritime pine was considered chemically closer to larch resin than resin from other pine species, although we now know that this is not accurate.

This raises a larger issue. Throughout the nineteenth century, chemists from all over the world studied resins in their laboratories, generating a vast number of experimental results. These results are, however, hard to interpret because the exact origin of the samples they used are not known. An influential French chemist, Claude-Louis Berthollet, was so discontent with contradictory results of his laboratory work on resin that he gathered resin himself from the trees to be sure of its origin. In fact, he confirmed the hypothesis that the turpentine spirit available on the market was usually a poorly distilled mixture of substances coming from different species [Dessort, 1869, p. 24].

Berthollet aside, many other great French chemists studied resin as well, for example, Louis Jacques Thénard, Jean-Baptiste Dumas, Eugène Soubeiran, Henri Deville (who wrote his dissertation on turpentine spirit), and Charles Gerhardt. Among these early pioneers, Auguste Laurent, the first chair of chemistry at the Bordeaux Faculty of Sciences in the 1830s [Brus and Menaut, 1938, p. 85], should also be mentioned. On the one hand, this interest shows that resins played a prominent role in the early years of organic chemistry. On the other hand, these early studies failed to construct anything resembling a proper discipline. This is due to their inconsistencies but also to the fact that chemists, while interested in final applications, for example, in the varnish industry, were less inclined to study the production processes of the forest industry. To understand the exact composition of resins, one would have to work with the producing industry, and not the consuming industry. It would be necessary to isolate a specific geographical area in a single climate with clearly identified botanical species and study the production processes, including resin harvesting methods and the techniques of small distilleries. Such a work would, by its very nature, be interdisciplinary and go beyond laboratory studies.

This kind of eclectic approach did not originate from influential university chemists. Instead, it was pursued by pharmacists from the Landes of Gascony 30 years before Vèzes. Turpentine was a traditional remedy in the region used against scabies, haemorrhoids, abscesses, boils, and infections of the womb and vulva, as well as for all kinds of respiratory diseases [Aufan and Thierry, 1990, p. 30]. In the nineteenth century, medical professionals expanded this list to include the treatment of typhoid fever, hydatids, taenia, rheumatism, lumbago, sciatica, and many more obscure diseases with turpentine. It was considered a true panacea, whose properties became cherished in countries with no natural pine forests [Dessort, 1869, p. 62].

Two local pharmacists constructed some of the most comprehensive and exhaustive works on pine resins, including detailed descriptions of resin harvesting and distillation methods, as well as properties and uses of local turpentine and rosin. The first pharmacist, Paul Dessort, born in Bordeaux, graduated from *École supérieure de pharmacie* (Higher School of Pharmacy) in Paris in 1869 with a dissertation *Du Pin Maritime et de ses produits* (On Maritime Pine and its products). It was divided into four chapters: botanical history, culture and product harvesting, chemical history, and pharmaceutical history. Three years later, in 1872, another student, this time from Gironde, Lucien Hazera, wrote a dissertation, *Etude du Pin maritime et des ses produits* (Study on Maritime Pine and its products). The second one is much shorter and it seems that Hazera was either unfamiliar with Dessort's work or he plagiarised it because the overall structure of Hazera's dissertation was very similar. Some of his insights were, however, original, and both works deserve to be read as an interdisciplinary introduction to the resin industry in the Landes during the nineteenth century and as a sign of things to come with the Pine Institute.

It is crucial to underline, however, that the pines themselves were studied by local foresters throughout the nineteenth century [Dive, 1864]. Although their works rarely dealt with the composition of pine resins, they are a valuable source of information on pine plantations, resin extraction methods, including distillation techniques, and the industry as a whole.

3. Between the Laboratory and the Forest: Contributions to Pine Gum Treatment

Maurice Vèzes, Georges Dupont, Georges Brus, and their collaborators understood that to avoid inconsistent results encountered in the works on resin from the nineteenth century, any serious research on pine resin chemistry had to be preceded by a careful study of harvesting and treatment methods. How exactly was resin harvested in the Landes? What were the variables that influenced the properties of resin? The history of harvesting, extraction and treatment methods is inextricably linked to resin chemistry, and both have to be studied together.

Harvest: Complex Simplicity

One of the earliest scientific descriptions of harvesting methods in the Landes region comes from Dessort's Ph.D. dissertation in 1869. He explained that pines started being tapped when they reached the age of 30–40 years.

> When a pine is old enough to be exploited, a *résinier* [forest worker specialised in resin extraction] standing on a type of ladder ... makes a 5 cm deep cut at a moderate height above the ground, around two meters. Every eight days, he enlarges the cut going up until it reaches about half of the tree's height. This wound gives resin all around a single campaign [Dessort, 1869, p. 13].

The campaign meant a period of exploitation in a given year, usually between March and October.

Hazera's account was slightly different compared to Dessort's. He wrote that the size of the *carre* ("face" in English) from where the resin is extracted should be 9 cm wide and 1 cm deep: "These dimensions give the best results preserving the tree at the same time." [Hazera, 1872, p. 10] He further explained that the wound should be refreshed every week and increased by 2 cm until it reached 60 cm by the end of the year. Finally, he explained that in winter, the *résiniers* would remove *barras*, the gum that hardened on the trunk, which could be used to manufacture low-quality products.

It seems that the authors of the works on pine resins had some discrepancies in their accounts of the depth of the wound (5 cm *vs.* 1 cm) and the duration of the exploitation of a single face. Dessort wrote that a new one had to be opened every single year (on the other side of the tree), while Hazera explained that a single face should be slowly increased over the period of five years. Hazera's observations are much more in line with the well-documented practices from the early twentieth century. The variability of practices within different parts of the same forest and the continuous evolution of techniques may explain the differences between the two accounts. While the works of Dessort, Hazera, and some of the local foresters often presented resin harvesting as an ancestral trade, a tremendous amount of expertise on how to exploit pines was improvised with the rapid extension of forestry areas in the nineteenth century. The locals were innovating and looking for new and more efficient techniques all the time — a topic explained in the works in the field of forestry from the period. Figure 15 illustrates the tools used by *résiniers* in the region.

In their resin chemistry manual *Résines et Térébenthines* from 1924, Vèzes and Dupont described the optimal way to extract gum:

> [T]he wound ... has the following dimensions: 9 centimetres large, 1 centimetre deep in sapwood, 4 centimetres of initial height. Every eight days, the résinier has to remove a light chip on the upper part of the wound in order to unblock ... the obstructed orifice of resin canals: this is the operation of piquage [chipping] (40 piquages a year). The wound has then a tendency to grow gradually; we continue the following years [Vèzes and Dupont, 1924, p. 86].

After extending the wound on the pine tree for a period of four years, it was recommended to allow the tree to rest for one year before opening another wound on the opposite side. It is evident that

Figure 15: French resin forest workers, *résiniers*. Left: 1864; Right: 1930s/1940s.
Source: Left: Samanos, 1864, p. 157; Right: Courtesy of Ecomusée de Marquèze-PNRLG.

a substantial part of the résinier's work revolved around the regular "chipping" of the tree to enlarge the wound and stimulate resin secretion. One forest worker was usually responsible for around 6,000 *carres*, each having to be chipped around 40 times a year (Figure 16). In total, he would conduct up to 24,000 chippings a year [Dupont, 1931] — this accounted for around 75 per cent of the activities of forest workers. In an attempt to alleviate this workload, Pine Institute researchers collaborated closely with the French Forest and Water Administration.

In 1931, a series of experiments took place in the Institute's experimental forest site in Pierroton in order to "reduce, at least partially, the chipping by preventing the face from drying up through harvesting the gum under a layer of water" [Dupont, 1933b, p. 74]. Special sealed bottles were used for this purpose, which were emptied with resin inside during harvest. However,

Figure 16: Landes maritime pines covered with *carres*. Date unknown but probably between 1930 and 1950.

Source: Courtesy of Ecomusée de Marquèze-PNRLG.

tests conducted between 1931 and 1933 concluded that the system did not work as the resin flow had to be manually renewed every week. The Pine Institute researchers tried to introduce various additives into the water: hydroquinone, sulfites, and acid solutions, but the results were disappointing, and the resin kept drying on the tree. Turpentine spirit was later used instead of water but also proved ineffective. The conclusion of the project was that chipping was an unavoidable part of the process of pine exploitation, and that chipping tools should perhaps be improved instead to make the task easier and faster. This conclusion was, as we will see, slightly premature.

The next step in the harvesting process after chipping is resin collection. Initially, the resin was gathered in a small cavity between the roots of the tree. This method was simple, but the resin was heavily polluted with organic debris [Dessort, 1869, p. 13]. However, this practice was questioned by Hector Serres, a pharmacist from Dax, France, who introduced small pots for resin collection on the tree. The technique was perfected in 1844 by Pierre Hugues from Tarnos (also in the Landes), who gave up his job as a lawyer and bought a small forest estate near the village of Boos to exploit it for resinous products. His technique consisted of installing small clay pots attached to the trunk just below the face (Figure 17). A small zinc gutter was fixed above to lead the flowing resin straight into the cup. Not only did the so-called Hugues system make it possible to obtain resin with much less forest debris, it was also richer in turpentine. Hugues' pots became popular in the 1860s (after Hugues' death), and by 1885, the previous method (collecting resin in cavities

Figure 17: Hugues' pot.

Source: Courtesy of Ecomusée de Marquèze-PNRLG.

between the roots) was almost completely abandoned [Sargos, 1931, p. 152; Vèzes and Dupont, 1924, p. 188].

In spite of its straightforwardness, Hugues' technique should not be dismissed as primitive. The exact shape of every component (clay pot, supporting nail, gutter) was designed in a way that made the installation fast and the setup solid. If forest workers had to spend too much time on fixing it, the entire device was fragile, or if its production costs were too high, the technology would not have gained a clear advantage over the older, cheaper and faster way of harvesting resin. Hugues' pots were an inspiration for similar equipment all over the world, including the United States (US), and they are still used in some forests in developing countries.

By the 1920s, the limitations of the clay pots became, however, increasingly apparent in France. Roger Sargos, a famous local forester and one of the early historians of the forest, explained the inconveniences in 1931: "[H]alf of the resin secreted by the tree is lost by overflowing from the pot ... and the gum loses by vaporisation ... more than one-third of its spirit; moreover, it oxidises and gets polluted by various debris." [Sargos, 1931b, p. 29] In response, the Institute sought new techniques to reduce installation time and better protect the gum from environmental factors. However, these two objectives were often at odds, as a more complex design would require greater skill to install and empty the pot after resin collection.

Techno-scientific Controversy over Resin Extraction: Bellini delle Stelle Case

A breakthrough was announced by an Italian engineer, Umberto Bellini delle Stelle, who moved to the Landes sometime in the 1920s. Originally from Florence, he, according to his own words, worked in Somalia, Eritrea, Malaysia, and in Brazil in rubber plantations [Sargos, 1931b, p. 29]. In 1927, he filed a patent for "the process and device for extracting gum, resin, and plant essences in general" [Bellini delle Stelle, 1929]. The *Proceedings of the French Academy of Agriculture* described the invention in the following way: "[W]ith a special tool having curved knives, like a milling cutter, and that is operated with a crank and gears, we make on the tree a circular concave orifice, the diameter of which starts at 0.05 m and reaches 0.1 m by the end of a [harvest] campaign" [Buffault, 1932, p. 90] (Figure 18). The wound had to be refreshed twice or thrice a week. The advantage of this method was that the face on the device was covered by a lid that protected it from air, sunlight and pollution. Bellini delle Stelle explained that the entire process would overall become less time-consuming, and the pine would be less exposed to diseases and insects than in the Hugues' system [Bellini delle Stelle, 1929, p. 2].

The new technique introduced by Bellini was not only aimed at protecting the wood but also at improving the quality of the resin. The resin gathered with Hugues' pots gave, after distillation, 20 per cent of turpentine spirit and 70 per cent of rosin (10 per cent of the weight was water, debris and adulterants). Bellini's apparatus, on

Figure 18: Bellini delle Stelle and his invention for extracting pine gum (April 1931).
Source: [Sargos, 1931, p. 7].

the other hand, was expected to provide no less than 32 per cent of turpentine, as it prevented evaporation of turpentine before the resin was gathered. The French were aware that similar technology had already been experimented with in the US a few years before, but Bellini's tool promised a much higher yield. In the early 1930s, when a great crisis struck the region, locals were desperately looking to science and technology to save the forest industry by increasing the quality and output of the resinous products. Surprisingly, their hopes were raised not by the institution funded by the industry for this purpose but by an outsider. And yet, despite the fact that Bellini delle Stelle took no less than nine patents between 1927 and 1933, his apparatus failed to achieve success. According to contemporary specialists in resin harvest, the device remained overly expensive and complicated to install [Frank, 2014]. It appears, though, that other factors also contributed to its lack of success.

The expectations for this technology were certainly very high. Bellini delle Stelle regularly featured in the pages of the local journal *Bois et Résineux* (Wood and Resinous Products), the largest weekly publication of the Landes industry, and Sargos himself hoped the new tool would revolutionise the Landes gum harvesting techniques.

In 1931, Dupont, the Pine Institute's director, officially endorsed Bellini's technique, and together with Sargos in 1932, decided to conduct a new series of experiments on the device's potential to replace the time-honoured Hugues' pots [Dupont, 1932, p. 51]. These experiments were funded by the new Landes Society for the Amelioration of Resin Harvesting (not to be confused with the Society of Studies and Applications for the Progress of the Resin Industry, which was an integral part of the Pine Institute). And yet, despite Dupont's early enthusiasm, the Institute's activity reports from 1932 completely ignored the project.

In fact, Bellini's apparatus raised suspicions very early on. In 1931, one of the numerous articles about this topic in *Bois et Résineux* explained that Bellini added a special anticoagulant in the receptacle that prevented the gum from drying — this substance was allegedly covered by a "secret patent". Indeed, in his early patents on different variants of the device, there are mentions of a secret anticoagulant that preserved the resin's fluidity. Why did the Italian inventor want to keep it secret? Sargos attempted to justify his behaviour: "[I]t was insinuated, even in the most official circles, that [the anticoagulant] was a 'humbug' ... it is not surprising that the inventor showed reserve towards us, the people of the Landes, not just because of our scepticism but because of our hostility." [Sargos, 1931b, p. 28]. However, in one of the later patents, the anticoagulant was defined as a mixture of turpentine spirit, terpineol, and aniline or ammonia [Bellini delle Stelle, 1931]. The doubts persisted because it was common knowledge among chemists over the period that gum did not coagulate if kept protected from air and water. In 1932, Dupont explained that adding an anticoagulant did not make any sense in a device such as Bellini's, whose purpose was precisely to limit the exposure to air.

In October 1931, Marcelle Barraud tempered the enthusiasm in the industry further: "Bellini's method has yet to confirm its efficacy," she explained [Barraud, 1931, p. 250]. But the tone of the debates degraded rapidly in the following months. In early 1932, Barraud emphasised that during economic difficulties, the people of the Landes should not trust too easily "promises such as the ones made recently by a Landes financial society", which had claimed that it was possible to conduct "on the tree itself chemical reactions and industrial processes" [Barraud, 1932]. This statement incensed Sargos, who believed that the society he was involved with was no different from the Pine Institute and was only funding research in the best interest of the local industry.

This led to a years-long debate between Dupont, Barraud, Sargos, and, of course, Bellini delle Stelle. Dupont tried to mediate and explain that the device may have been of great merit, but Barraud could not be accused of crossing the line because she gave her informed opinion on her topic of speciality [N. N., 1932c, p. 266]. Barraud openly called many of Bellini's claims "anti-scientific". According to the Italian engineer, his device made it possible to gather something called "ephemeral gum", and its distillation would make it possible to obtain substances, such as terpineol, terpine, terpinolene, and dipentenes, which were usually available only through the transformation of turpentine spirit. Not only that, he wrote in his correspondence that the terpine mixed with the anticoagulant would turn into terpineol and various terpenes in a receptacle attached to the tree. The chemical reactions were supposed to be accelerated thanks to the movement of the recipient induced by the wind.

Dupont did not believe that any of these products could be manufactured outside of chemical plants though. In his reply to Bellini delle Stelle, he tried to remain diplomatic by maintaining that the device might be useful but warned Bellini that "it is dangerous to promise [the production] of terpine, terpineol, as well as of borneol and camphor, before obtaining the results". Bellini was not, however, taken aback by this criticism and wrote to Dupont: "I reaffirm my claims, even if you continue to consider them dangerous." [N. N., 1932c, p. 266] It is clear that Dupont had a hard time positioning himself in this case. On the one hand, if Bellini's invention could increase even a little bit the global annual yield of the resin industry, it was worth trying. On the other hand, he could not accept the pseudo-scientific affirmations in his own field of expertise, chemistry. He decided to answer via the *Bulletin de l'Institut du Pin* three big questions often addressed to the Institute about the new method of Bellini: "1. What do you think about the anticoagulant? 2. What do you think about the ephemeral gum? 3. Do you believe we can extract more valuable products, such as camphor, resin ethers, terpine, terpineol or others, from the recipient?" [N. N., 1932c, p. 269]

The first question was the simplest to respond to. Regarding the anticoagulant, Dupont repeated his stance that it was unnecessary in the absence of air during gum collection. Bellini's notion of "ephemeral gum" was also refuted, as Dupont explained that it existed but did not transform into regular resin due to exposure to air. Instead, it underwent a biochemical process while leaving the resin-producing ducts in the sapwood, which meant it could not be harvested like regular resin. In terms of composition, there was nothing distinctive about the resin collected using Bellini's device. As for the last question, Dupont explained that the only product that could be manufactured in the receptacle was terpine if, for example, sulfuric acid was added, but the yield would be weak and the separation difficult. In other words, all of Bellini's claims were scientifically unfounded.

Although it is perfectly possible that his device was too costly and unpractical, the fact that no single mention of Bellini delle Stelle appeared in the Pine Institute Bulletin and in *Bois et Résineux* after 1933 may suggest that the immediate cause for abandoning this technological promise was the heated conflict between him and the local scientific experts.

Introduction of Resin Harvesting with Acid

The polemic between Bellini and the Pine Institute researchers did not mark the end of attempts to improve harvesting techniques. In 1934, the Pine Institute researchers experimented with a new and improved design of Hugues' pots with lids to limit pollution with bark, needles and water. Again, while the innovation may have seemed straightforward, the practicalities were not. The lid should not make the process of emptying the pots into barrels slower or fall into the pot itself. The Institute's chemists spent days with forest workers in the forest itself, observing their gestures and the way the workers interacted with their tools in order to identify the optimal design of the new model of the lid-covered pot. In the end, the project was abandoned due to the lids' fragility (they notoriously fell into the pot, breaking in the process) and the time-consuming way of their setup [Brus, 1935b, p. 42].

In 1936, a novel technique captivated the attention of local scientists and industry experts: *le gemmage activé* ("gum harvesting with acid"). The method was invented in Germany and involved spraying the pine's wound with a solution of sulfuric acid to increase gum secretion. The Germans hoped that this method would increase the total yield of turpentine, enabling them to produce terpenic products on their own territory without depending on French and American imports [Hessenland, 1935, p. 636]. Their experiments were, however, conducted on a different species (*Pinus sylvestris* and not *Pinus palustris*) and in an industry considerably less developed.

When the Pine Institute researchers replicated these studies, the results were disappointing. The yield only slightly increased when chipping was conducted every two weeks. In Aquitaine, however, chipping was traditionally performed every week — a practice to extract more resin (the frequency of chipping was typically conditioned by the available workforce). With such an interval, this technique of adding sulfuric acid was "completely devoid of interest in our region", according to the Institute's report for 1936 [Brus, 1937, p. 64].

The reasons for the Pine Institute's abandonment of the acid spray technique in 1936 are difficult to assess in hindsight, as are the reasons for the failure of the experiments. Nevertheless, ten years later, in 1947, the Pine Institute researchers were surprised by the advancements of this technique in the US. After a French delegation's visit to the US that year, the American method of using acid spray for resin extraction was discussed widely in the regional press [Prat, 1947]. The Pine Institute and the local forestry services decided to implement this technology on French soil. This time, the experiments were much more promising [N. N., 1948]. By 1952, Charles Prat, one of the local industry leaders, announced that the industry was on the verge of a revolution as important as the introduction of Hugues' pots a century before [N. N., 1952]. In 1955, Georges Brus reported that after seven years of study, it was beyond doubt that the introduction of acid could increase the yield by 25 per cent.[51] In 1950, 20 per cent of all resin harvested in the US was obtained through acid-based harvesting. By 1955, this proportion had increased to 60 per cent [Brus and Legendre, 1956, p. 8]. France seemed to be following a similar path. However, in a surprising turn of events, the Annual Congress of the Forest Workers in 1955 rejected the new technique, and was supported by the French Communist Party, which believed the technique would increase forest owners' profits at the expense of the safety of forest

[51] Gironde Departmental Archives, Pine Institute Archives, box "Fréjacques". "Compte rendu de la deuxième rencontre à Bordeaux de la Commission Fréjacques", 30 June 1955.

workers. The process of covering the incision on the pine with a 50 per cent sulfuric acid solution was seen as dangerous. Moreover, the official position of the Congress was that the acid damaged the tree permanently, potentially threatening the survival of the forest and of the entire local forest-based industry [Frie, 1956, p. 48]. The Pine Institute's laboratory of forest biology conducted additional studies to reassure the locals that the technique was not dangerous to the trees. Its researchers also launched a campaign in the pages of the local press to persuade the *résiniers* that, in spite of inconveniences, the economic benefits would be indisputable. The Institute's involvement bore fruit despite opposition, and by 1960, 50 per cent of the resin in the Landes Forest was extracted with acid [N. N., 1964].

Gum Distillation Techniques

Once a forest worker poured resin from a pot into a barrel, it was then transported to a distillery for further processing. The subsequent step in the process involved cleaning the gum and separating it from any debris and water. This was achieved by heating it to approximately 50–60 °C and adding around 10 per cent of turpentine spirit, making the resin more fluid and enabling it to be filtered. This process was called "*térébenthinage*". Such a purified resin was ready for distillation.

Between 1852 and 1922, no less than 50 patents were taken out in France on resin purification and distillation. It is not surprising that these operations were considered a priority for the Institute's chemists. Throughout his life, Vèzes published at least twelve articles describing different industrial methods used in the region and suggesting their improvement.[52]

How does one distil pine resin? Brus and Legendre explained the process in detail in one of the last great general introductions to resin

[52] The list of publications authored by M. Vèzes, *Bulletin de l'Institut du Pin*, 2, 1935, p. 31–35.

chemistry produced by the Pine Institute, an article in the encyclopaedia of paints and varnishes from 1956: "The separation of turpentine spirit cannot be done by a simple distillation under atmospheric pressure because it would require a temperature so high that it would cause the resin acids to decompose. It has to be done, either in vacuum, in a temperature inferior to 180°C, or by agitation with water vapours." [Brus and Legendre, 1956, p. 11]

Resin distillation in the Landes region underwent several stages of development over time. Initially, in the eighteenth century, the process involved using naked flame alembics, which were heated using firewood and required the gradual addition of water during the distillation. Later, water was replaced by steam coming from a boiler. In the next phase, the alembics ceded their place to more sophisticated designs with "bundled tubes". However, all of these methods relied on discontinuous distillation, which meant that it was impossible to extract all the turpentine from the resin. Consequently, the Pine Institute began advocating for the introduction of continuous distillation on a larger scale in the 1920s [Dupont, 1926, p. 55].

The oldest continuous distillation device was built in 1909 by the local company Castet. It was not only continuous but also operated under a vacuum, producing high-quality final products [Vèzes and Dupont, 1924, p. 242]. Despite its advantages, Dupont noted that only large companies could afford it and that it did not distil thoroughly enough, leaving between 1.5 per cent and 3 per cent of residual turpentine spirit that could have been extracted. He aimed to develop a more affordable machine with similar or better-quality output [Dupont, 1926, p. 56]. In 1924, Dupont filed a patent on what was probably his most important contribution to the industry: the distillation column using steam (Figure 19). By 1927, the first models started being sold by the Institute to local companies [Barraud, 1927, p. 81].

In the subsequent ten years, a local inventor from Bordeaux, Emile Courret, closely collaborated with the Institute to enhance the apparatus. In total, he took out three patents on different improvements of the column [Courret, 1929; 1934; 1940]. The Institute

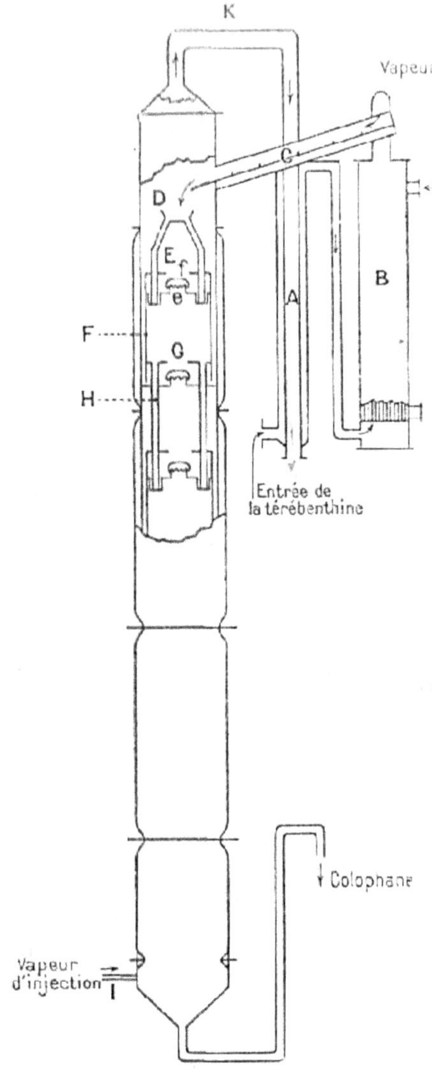

Fig. 20. — Appareil continu à vapeur.

Figure 19: Dupont's continuous steam distillation device.
Source: [Dupont, 1926, p. 57].

continued to promote, advertise and sell the apparatus, but it did not directly compete with Castet, whose distillation column was also recommended and even improved by the Institute's researchers with the company's engineers [Vèzes, 1932].[53] For Dupont and later Brus, the goal was not to earn money by selling their inventions but to bring lasting changes to the practices of the local industry.

And yet, the true impact of both columns is rather ambiguous. In 1956, 65 out of 80 resin distillation plants in the Landes, some of them huge, still used discontinuous distillation. Eight used the Pine Institute's column, and seven used Castet's column [Brus and Legendre, 1956, p. 11]. The reasons for this outcome were numerous. While the continuous distillation required less supervision, consumed less energy, and gave better quality products, the continuous distillation columns were also considerably more expensive, much more fragile (needing more repairs), and their installation required well-trained personnel. Overall, the vast majority of small workshops could not simply afford them.

However, this may be also have been due to the fact that the quest for more efficient distillation methods ceased to be at the heart of the Institute's preoccupations in the 1940s. A new perspective opened up: the fractionation of resins into products much better defined than turpentine or rosin. For Brus, it was evident that the contemporaneous methods were highly deficient. The huge variation in the exact composition and properties of raw materials made it difficult to define the optimal conditions for the manufacture of derivatives.[54] This echoes the debates in the nineteenth century on the composition of turpentine. The consuming industries had to deal with the fact that turpentine and rosin coming in different deliveries throughout the year might have different properties. This was a threat for the resin producers because the consumers, such as the paints and varnishes industry, could easily turn to substitutes coming from petroleum fractionation, which might have been of lower but more consistent quality. In 1950, the Pine Institute announced that after many years of research, it had developed a new mode of gum fractionation and

[53] See also: Gironde Departmental Archives, Pine Institute Archives, classeur "Lettres techniques 1952, F". Letter to Food and Agriculture Organisation of the United Nations, 16 May 1952.

[54] Gironde Departmental Archives, Pine Institute Archives, file "Lettres Techniques 1952–1953, G". Letter from the Pine Institute to J. Gillet, 31 October 1952.

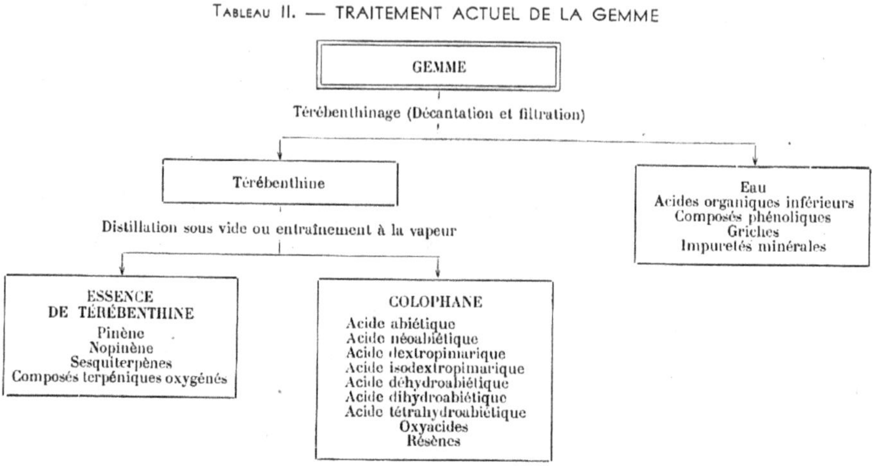

Figure 20: Traditional gum treatment process (Mode 1). The gum (on the top) can be transformed into clean gum (*térébenthine* — middle left) and waste (middle right). Clean gum can be distilled into turpentine spirit (*essence de térébenthine* — lower left) and rosin (*colophane* — lower right).

Source: [Brus, Lê Van Thoi *et al.*, 1952, p. 866].

filed two patents [Brus and Legendre, 1951; 1951b].[55] Brus compared the old and the new mode. For him, the old one, practiced for more than one hundred years, consisted of two steps: cleaning (*térébenthinage*) and distillation into two products, turpentine spirit (*essence de térébenthine*) and rosin (*colophane*), both of which could be later transformed or fractionated into other final products (Figure 20).

The new mode was considerably more complex. It consisted of two phases: the treatment of the gum in the first phase and the fractioning of different acids present in rosin in the second phase. Unlike the old mode, the new mode did not require distillation but instead used a range of more targeted processes (Figure 21). The resin was first diluted with a mineral essence to facilitate filtration and cleaning without the need for heating. Next, the resin acids were treated

[55] Gironde Departmental Archives, Pine Institute Archives, box "Divers". Brus, G. "Travaux récents effectués à l'Institut du Pin, discours présenté au Syndicat des Sylviculteurs", April 1950.

TABLEAU III. — NOUVEAU TRAITEMENT DE LA GEMME

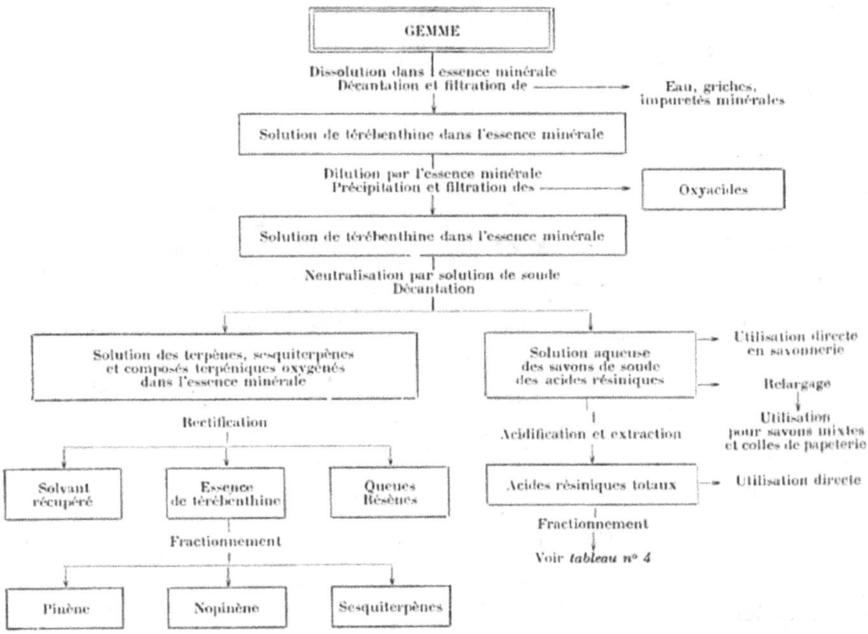

Figure 21: New gum treatment process (Mode 2, Phase 1: gum fractionation). Gum (top) is purified in a two-level process. Then, by neutralisation with sodium solution, it gives resin soaps (*savons aqueses* — middle-right) and a terpene solution (*solution des terpènes* — middle left). Both can be further rectified and fractionated (bottom).

Source: [Brus, Lê Van Thoi *et al.*, 1952, p. 866].

with a sodium solution to separate them from the solution of terpenic compounds by decanting. The solvent was then recovered, and the solution was fractionated into desired terpenes. The resinous soaps could either be used directly in soap or papermaking or acidified to pass to the second phase of rosin treatment to separate individual acids (Figure 22).

The researchers at the Pine Institute were meticulous in their examination of the potential uses for each byproduct resulting from the new fractionation method. They recognised that the oxyacids, which made up only 2 per cent of the overall weight and were

TABLEAU IV. — FRACTIONNEMENT DES ACIDES TOTAUX

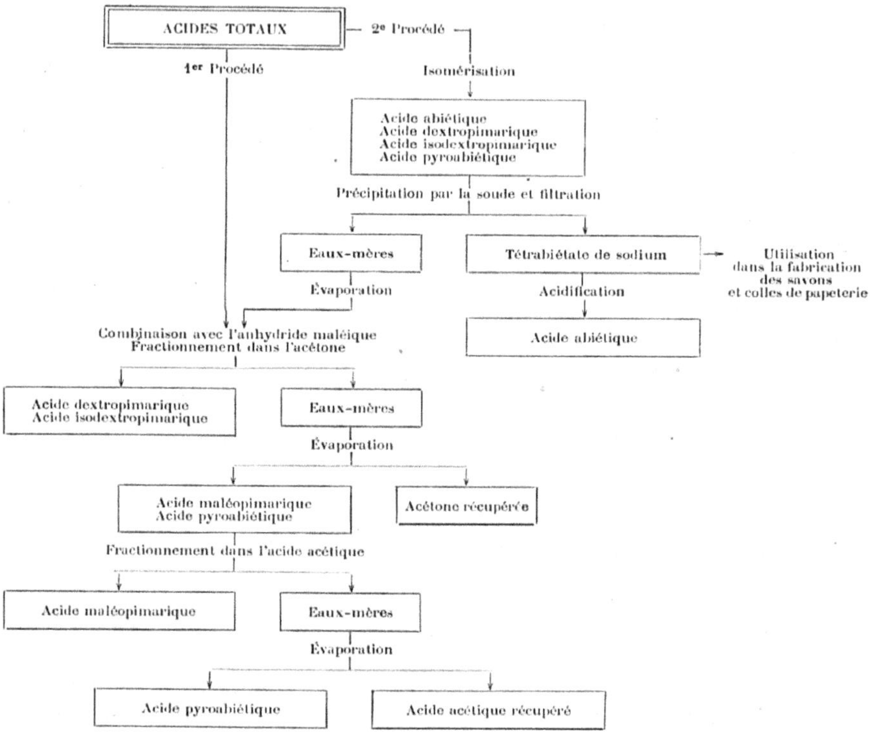

Figure 22: New gum treatment process (Mode 2, Phase 2: resin acids fractionation). A multi-level treatment of resinic acids obtained from the soap produced in the previous stage.

Source: [Brus, Lê Van Thoi *et al.*, 1952, p. 867].

separated in the first phase, could still have practical applications in varnish-making and electric insulation.

The novelty of the process lay in the fact that it was possible to obtain many previously unavailable acids in the second phase due to the lack of thermal treatment. The Pine Institute considered maleopimaric acid the most promising product, obtained through the reaction between levopimaric acid and maleic anhydride. This acid generated a lot of interest within the Institute, which worked on exploring its potential applications and derivatives for several years [Brus, Lê Van Thoi *et al.*, 1952].

The main difficulty with the new mode of gum fractionation was its complexity and the very high cost of new installations. As a consequence, from 1950 on, the Institute worked on simplifying it and finding applications for the newly discovered acids to show their superiority over petroleum derivatives.[56] Furthermore, the Institute had to convince the local industry of the potential of the new method in comparison to older valorisation strategies. In a speech at a forestry congress in 1950, Brus explained that this new method was nothing but the next step in a very ancient history of resin uses: "At first used in a raw state ... gum became by the end of the eighteenth century distilled in order separate its constituents ... Isn't it natural to envisage a more selective fractioning?"[57] Brus argued that the virtues of the new method were already acknowledged in the US and by the petroleum industry, that "does not content itself any more with the fractioning of petroleum into spirits, kerosene, gas oil, oils and pitches, but it isolates pure hydrocarbons ... The petroleum industry is becoming a powerful chemical industry, and it would be unwise to underestimate what its products will do to the resinous ones".[58]

The threat of oil loomed large in the region, and Brus underlined that clients would become increasingly selective. He argued that without fractionation, the industry would obtain mediocre products, which would "have no commercial future". Was this speech necessary altogether? Wasn't the necessity to innovate self-evident? What is important to underline is that the local industrial ecosystem

---

[56] Gironde Departmental Archives, Pine Institute Archives, box "Divers". G. Brus, "Résumé des travaux effectués par le laboratoire de M. Legendre au cours de l'année 1951", July 1952.

[57] Gironde Departmental Archives, Pine Institute Archives, box "Divers". G. Brus, "Travaux récents effectués à l'Institut du Pin, discours présenté au Syndicat des Sylviculteurs", April 1950.

[58] Gironde Departmental Archives, Pine Institute Archives, box "Divers". G. Brus, "Travaux récents effectués à l'Institut du Pin, discours présenté au Syndicat des Sylviculteurs". April 1950.

remained profoundly divided, with tens of companies occupying different niches in the resin treatment cycle. On top of that, there were conflicted forest owners' associations and forest workers' unions. The entire industry was in a delicate equilibrium that a radical technological change could threaten. As explained, the forest workers opposed gum harvesting acid, expecting it might lead to the reduction of jobs in the region. Similarly, the introduction of resin fractioning, due to the high costs, would require many mergers — an unwelcome evolution for many entrepreneurs.

Some economic misjudgements made by the Institute should also be mentioned. In fact, the Institute's chemists had a harder time finding applications for the much-advertised maleopimaric acid than Brus had initially expected. In 1956, he expressed disappointment that there was still no manufacture on the industrial scale in the region [N. N., 1958]. Around 1962, the acid was integrated into the papermaking process, but it remained a marginal outlet. In 1968, some semi-industrial experiments were conducted on maleopimaric acid derivatives, but most of the local producers synthesised the acid from rosin and not by gum fractioning.

In 1971, one of the Institute's engineers advocated anew for the introduction of fractionation as if there had been absolutely no progress in that matter over the previous twenty years.[59] Despite the Institute's desire to emulate the developments in the petroleum industry, the local economic environment did not support this direction.

4. Turpentine Spirit: Study of its Properties and Derivatives

After harvesting, distillation and fractionation, pine resin is usually processed into the volatile turpentine and solid rosin. Turpentine spirit was, until the middle of the twentieth century, the main and most valuable pine resin derivative. Like pine resin, turpentine spirit

[59] Gironde Departmental Archives, Pine Institute Archives, folder "Saint Symphorien". R. Bentejac, "La gemme, sa production, son utilisation", 1971.

was a complex substance whose physical and chemical properties were conditioned by a plethora of factors, and the product available on the market was far from uniform. As a consequence, tradespeople were interested in understanding its properties to describe the turpentine they were distributing accurately, detect fraudulent practices, and identify counterfeit products. These commercial motivations aligned with scientists' interests in better understanding the substance's exact composition, as they were dissatisfied with the confusing results obtained in the previous century.

Between Science and Commerce: Identifying Properties of Turpentine Spirit

In the 1900s, turpentine spirit was the most important research topic for the Laboratory of Chemistry Applied to the Resin Industry. More than half of all scientific articles published between 1900 and 1920 by Vèzes were about turpentine spirit. What were these early studies about? Whatever the method used to distill resin, turpentine spirit always contained some proportion of polluting agents. Vèzes worked on detecting them [Vèzes and Mouline, 1901] and the refractometry of these adulterants [Vèzes and Mouline, 1901b] as well as on their composition [Vèzes and Mouline, 1902].

In general, the local researchers differentiated normal and unusual adulterants. The first resulted from bad distillation or from excessive oxidation. They included rosins (resin acids), oils and oxides, and they were tolerated to a certain degree because even in the best stills, some level of pollution was unavoidable. The second category was the "impurities" introduced explicitly with dishonest intention. It was in 1904 when Vèzes mentioned for the first time, that the number of samples of turpentine diluted with white spirit were particularly high that year [Vèzes, 1905]. What is white spirit? Today, we know it is a mixture of aliphatic, open-chain or alicyclic C_7 to C_{12} hydrocarbons, with a boiling range between 130 °C and 200 °C, although the exact chemical composition varies depending on multiple factors. For an untrained person, the addition of white spirit to turpentine is hard to detect since the two liquids share many properties, but the white spirit's quality as a solvent is significantly inferior [Beauclair-Lafaye, 1914].

In 1905, Vèzes proudly announced that his laboratory was capable of identifying the adulteration of turpentine with mineral spirits with 100 per cent accuracy [Vèzes, 1905]. The public authorities were aware of the growing scale of frauds, and in an official ministerial decree in the same year, they entrusted Vèzes' laboratory with the task of analysing turpentine samples from the entire French territory. From that moment on, the research results obtained by Bordeaux chemists established a benchmark for the quality of the products on the French resin market and gained a binding power in litigations.

The authority of the resin laboratory grew over the years, and in the 1920s, the Institute, under the direction of Dupont, played a significant role in setting trade standards for turpentine. It carried out research to define clear-cut brackets of the properties that a well-distilled turpentine spirit should satisfy. The nature of the Institute's research on turpentine spirit can be illustrated with Barraud's Ph.D. dissertation [1927], which continued to be a reference in the field at least until the 1950s.[60]

In the 1920s, it was generally accepted that the maritime pine's turpentine spirit started boiling at 153 °C, that 80 per cent of the essence was distilled below 164 °C, and that its density, at 15 °C, was between 0.855 and 0.865 (although this range was debated later). As for the chemical composition, it was discovered that the turpentine was mostly made of two terpenes: levorotatory α-pinene (63 per cent) and dextrorotatory β-pinene (25.5 per cent), also called nopinene. The rest (about 10.5 per cent) was made of sesquiterpenes, resens and oxides [Dupont, 1926, p. 78]. The exact proportions continued to be redefined, corrected, and adjusted for decades. In the 1950s, the average proportion of α-pinene was evaluated to be 67 per cent [Brus, Legendre and Niolle, 1952]. But it must be

[60]Gironde Departmental Archives, Pine Institute Archives, file "Correspondance 1952, D". Letter from Darrade Frères, 7 May 1952.

reiterated that these were average values that concerned only maritime pines from the Landes Forest. Barraud went one step further and prepared a larger comparative study.

The title of Barraud's work was *Contribution à l'étude des essences de térébenthine* (Contribution to the study of turpentine spirits), and it was divided into three chapters. The first chapter was an in-depth comparison of the turpentine from Austrian pines and maritime pines. Barraud studied the variability of the yield and the turpentine's properties, which depended on the harvest season and dimensions of the incision, as well as meteorological conditions. Her primary interest was the optical rotation of turpentines. For Dupont, it was obvious that optical rotation was a key property in detecting frauds, since white spirit had no optical rotation, and its addition to turpentine spirit would sensibly reduce the overall optical rotation of the mixture [Dupont, 1926, p. 258]. The necessary condition for a test to work was to know the origins of the sample because turpentines from different species had different terpenes and, as a consequence, variable optical rotation.

Barraud's work nuanced this approach. She demonstrated that the optical rotation varied greatly inside the same species between individual trees. It was accepted over the period that the average rotation of turpentine was –31 (the sodium D line), which was more or less correct for commercial mixtures coming from many different trees. However, in reality, the value could range from +4 to –40 between different harvests from individual trees [Barraud, 1927b, p. 25]. Thus, fraud detection was a complex process that required multiple analytical methods used concurrently. In the second chapter of her dissertation, she provided an overview of the properties of turpentines from various pine species, with an emphasis on optical rotation.

The last chapter presented a wide range of analytical methods used by chemists to analyse turpentine. Barraud underlined that chemists often get confused by the sheer number of potential tools, and her goal was to clarify their uses [Barraud, 1927b, p. 60]. This chapter highlights one of the primary characteristics of the type of scientific work conducted in the Institute. Barraud gathered and summarised research in the field and added her own observations, but more importantly, she tried to construct a more theoretical overview of the practical challenges encountered in the laboratories of resin chemistry. Barraud's research, while immediately useful for the industry, was also relevant for the purely disinterested pursuit of knowledge on the properties of turpentine in general.

The development of fast and reliable methods for detecting fraud was crucial in the studies on turpentine spirit. However, counterfeit turpentines remained only a part of a much broader problem. Of course, fraudulent practices were forbidden, but the use of substances like white spirit was not. White spirit became a very popular solvent in the paint and varnish industry in the first decades of the twentieth century — its price was significantly lower than turpentine because of its lower quality. The Pine Institute had to react to its rising popularity in order to help the local resin industry. To achieve this, two tasks had to be undertaken: first, convincing potential customers that the quality of turpentine spirit justified the price, and second, finding new market outlets for turpentine.

War Against Substitutes: White Spirit and Foreign Turpentine

Why was turpentine a better choice than white spirit? First, the characteristics of turpentine spirit were, in general, relatively well-established in the 1920s, unlike those of white spirit, which was a mixture of light and heavy products [Gérard, 1931]. However, the more important difference in terms of quality was its siccativity. White spirit was an inert solvent that volatilised during the drying process. As a result, paints and varnishes that relied on white spirit required the addition of metallic drying agents, which could compromise their quality and durability [Gérard, 1931]. Turpentine spirit was, on the contrary, an excellent drying agent on its own. Barraud showed in her research that turpentine was an active solvent because it fixed air oxygen and then transmitted it to line oil, one of the key components of many vanishes. Thanks to this property, it made paints and varnishes dry faster and increased their hardness [Barraud, 1930, p. 147]. Additionally, turpentine had wetting and dispersive properties, which made it easier to mix with paint and gave it a certain level of elasticity. The Pine Institute Bulletin published many articles that presented the advantages of turpentine over mineral solvents and that argued for their wider use; see [Dupont, 1931; Dupont, 1933; Snitter, 1933].

Despite the numerous advantages of turpentine over white spirit, the fluctuating prices of turpentine in the 1920s and 1930s posed a significant challenge for the industry; for example, in 1926, 1 kg of turpentine cost 10 francs, while 1 kg of white spirit was 1.8 francs, more than five times less. Even after a significant drop in prices following the resin market crisis, the difference in prices remained considerable, with turpentine priced at 3.5 francs and white spirit at 2.75 francs [Barraud, 1930]. In other words, even in the middle of the worst crisis in the history of the resin industry, turpentine was more expensive than its main competitor. Was this saving reasonable for the varnish industry, considering the low quality of white spirit? The difficulty was that all the studies of the Pine Institute on turpentine spirit and its promotional activities concerned the so-called "fat turpentine", meaning an already partly oxidised turpentine spirit. Fresh turpentine spirit would evaporate quickly and entirely from a varnish, while the fat turpentine formed a thick film, giving the varnish elasticity and, as already explained, it facilitated the oxidation of oils [Dupont, 1932, p. 57]. Paradoxically, this problem did not occur before World War I, when the distilleries stored turpentine for many months before selling it. This led to a partial oxidation of turpentine in primitive containers. After the war, because of the fluctuations in the resin prices, the workshops tried to sell turpentine as fast as possible, and the varnish producers often got a "too fresh" turpentine spirit of inferior quality.

This challenge preoccupied the Pine Institute for many years. In 1927, its chemists attempted to accelerate the oxidation of pinene using different catalysts. They discovered that cobalt acetate significantly increased the rate of oxidation [Dupont, 1929, p. 151]. In 1930, the Institute established a semi-industrial facility for the production of fat turpentine, but it is unclear whether the initiative was successful, as the Institute's activity reports do not mention the outcome. Nevertheless, the Institute actively promoted fat turpentine spirit throughout the 1930s and collaborated with the state-funded Aeronautical Research Service to establish a workshop to investigate the quality and durability of paints and varnishes used on aircraft [Dupont, 1933b, p. 75]. The result indicated that varnishes based on fat turpentine were more resistant to harsh weather and chemical agents [Snitter, 1933, p. 225].

Apart from competition from white spirit, the French turpentine spirit also faced competition from imported turpentine, particularly from the US. The superiority of French turpentine was widely acknowledged before World War I [Gildemeister and Hoffman, 1900, p. 263]. After the war, however, it was not the case any more, notably because of the colouration of the French turpentine in commerce that, unlike the American one, was not discoloured after the distillation. The Pine Institute conducted research on the causes of discolouration and identified four reasons. Firstly, turpentine spirit was yellow if it was too old or contained too much rosin due to bad distillation. Secondly, it was reddish if it was polluted with iron from barrels. Thirdly, it was green if it was polluted with copper salts. Finally, it was opaque if it contained too much water [Dupont, 1927]. Over the period, two discolouration techniques were known: using sulfuric acid and caustic soda. However, in both cases, the byproducts were hard to remove. To address this issue, the Pine Institute developed a new technique for filtering turpentine with solid oxalic acid, which effectively absorbed metals and water. This technique continued to be recommended throughout the years, but in 1933, Barraud wrote that the most effective way to prevent certain forms of discolouration was to paint the interior of barrels and containers with a varnish or lacquer that was not soluble in turpentine spirit [Barraud, 1933, p. 132]. Despite these efforts, the problem of discolouration persisted, and the Institute had to regularly answer questions from concerned producers on how to avoid it. In 1953, in a brochure distributed to local companies, the Institute recommended cleaning the barrels from rust, applying a turpentine-resistant varnish inside them, and if the prevention failed, the brochure advised to use the discolouration by oxalic acid, developed 30 years prior.[61]

Synthetic Camphor: A New Hope for the Landes Industry

Aware that the competition from white spirit and other substitutes was a formidable challenge, the Pine Institute explored new potential uses for turpentine. As Dupont observed in 1926 [p. 271], the chemical applications of turpentine were still in their infancy, partly because the complex composition of turpentine spirits was poorly understood and also because the industrial yield of chemical reactions involving turpentine was too low to be economically viable. Despite these obstacles, the Institute persisted in its efforts to find innovative ways to harness the unique properties of turpentine.

One of such innovative outlets was the manufacture of synthetic camphor, a cyclic monoterpene ketone with the chemical formula $C_{10}H_{16}O$. Natural camphor was extracted through distillation from camphor trees found in China, Japan and Taiwan. By the end of the nineteenth century, the camphor industry in China was still in the very early stages of development and in Japan, it was already declining due to the disappearance of camphor trees. It was Taiwan, or the Isle of Formosa back then, which was at the heart of the global camphor extraction industry. Camphor was of great economic importance because it formed with nitrocellulose, a colloidal solution of remarkable plastic properties, called celluloid. The first celluloid-like material was invented in the 1860s by a British inventor, Alexander Parkes, and later independently by an American, John Weasley Hyatt, who, according to popular story, sought to replace ivory in snooker balls [Bijker, 1987]. Celluloid was quickly adopted in other industries, including the manufacture of tennis balls, toys, dolls, piano keys, and explosives. Nevertheless, the true revolution came in 1888 when celluloid became indispensable for the photographic industry.

---

[61] Gironde Departmental Archives, Pine Institute Archives, box "G 30". "Sur quelques défauts des produits résineux et les moyens de les éviter", 22 July 1953.

In 1895, Taiwan became the first colony of the Japanese empire. By 1900, the profits from the camphor trade accounted for no less than 27 per cent of all total net tax revenues of the empire, and in the following years, the camphor monopoly continued to generate more than the 10 per cent of total net tax revenues in Japan [Feng-en Tu, 2015]. Many countries sought to become independent from the Japanese supply and tried to produce camphor on their own. The challenge was that camphor could only be extracted from trees that were at least 60 years old, with the best yield from those over 100 years old [Brus, 1934]. Consequently, plantation projects initiated by several western nations in the early years of the century had to wait for several decades before becoming productive. It is not surprising that many turned to chemistry for assistance.

Synthetic camphor was first manufactured, long before its glory days, by the influential French chemist Marcellin Berthelot in 1858.[62] But it was not until the end of the nineteenth century that Finland and the US established the first synthetic camphor production plants [Feng-en Tu, 2015]. The rapid increase in camphor prices during and after the Russo-Japanese War in 1904–1905 made synthetic manufacture an economically justifiable investment. Between 1907 and 1911, many synthetic camphor-producing companies were established in France, England and Germany [Brus, 1934]. The Japanese reacted quickly to this new rising competition by reducing the price of natural camphor and leading its competitors to bankruptcy, with the exception of the German company Schering. After World War I, Japan continued to profit from its monopoly and imposed exorbitant prices. However, the over-exploitation of Taiwanese forests had left the Japanese with little room for manoeuvre, creating favourable conditions for the synthetic camphor industry to re-emerge in Europe [Dupont, 1926, p. 271]. The Landes Forest could potentially benefit from the forest crisis on the other

[62] J.-M. Michel, “Les problèmes du celluloïd”, Contribution à l’histoire industrielle des polymères en France, p. 3 (www.societechimiquedefrance.fr/IMG/pdf/a_1_324_100.vfx2_sav.pdf, last accessed 18 February 2021).

side of the world by synthesising camphor from pinene in turpentine spirit. Dupont envisioned that the French celluloid industry, in fact, could take advantage of the proximity of pine forests instead of relying on foreign imports [Dupont, 1926, p. 273]. However, the industrial production of camphor was problematic not only for economic reasons (with Japan still being capable of influencing the prices) but because the industrial synthesis methods were relatively complex and the yield of camphor was no higher than 60 per cent of the turpentine spirit used in the process.

The first works by the Pine Institute on synthetic camphor were conducted in 1924 and 1925 in order to discover and understand the methods used abroad [N. N., 1925, p. 3]. According to Dupont and Brus, the industry in Aquitaine should have adopted the method of Schering, which, despite its complexity, had a reasonable yield [Brus, 1934]. It did not mean that the Bordeaux chemists wanted to passively copy the German process. On the contrary, Dupont and Brus (still in Toulouse back then) filed a patent on the "Perfecting of the synthetic camphor manufacture" in 1925, which went beyond the Schering method [N. N., 1925b, p. 272].

According to Dupont, the construction of a new industry in France was imminent in the middle of the 1920s, and in his publications, he gave a great deal of technical advice on how to make it happen. He stressed the importance of constructing production plants near the distilleries to minimise transportation and storage time. This was because the camphor industry required turpentine that was not oxidised, unlike the paint and varnish industry [Dupont, 1926, p. 276]. In other words, he advocated not only for the French camphor resin industry but specifically for the Landes one. In 1928, Dupont oversaw the construction of the first semi-industrial plant in the region, which was scheduled to begin operations in 1930, but the project was ultimately abandoned due to new fluctuations in camphor prices [N. N., 1928].

In 1934, the new director of the Pine Intitute, Brus, decided to rekindle the interest of resin producers in the potential of the camphor industry:

> Currently, the synthetic camphor industry does not exist in France, and we find ourselves in a situation which is both paradoxical and distressing, to see our country, the first European producer of turpentine spirit, having to buy abroad the camphor necessary for its factories, for its powder production plants, while the German and Italian companies, which have to import turpentine spirit, manufacture not only the camphor necessary for their consumption but can also sell this product on foreign markets [Brus, 1934, p. 108].

In 1933, France consumed around 800 tons of camphor: 66 per cent for celluloid, 12 per cent for pharmaceutics, and 10 per cent for explosives. Half of it was imported from Japan, and half from Germany and Italy, the two countries with important synthetic camphor plants. Interestingly, it seems that it was precisely the abundance of raw materials that slowed down the development of the French camphor industry, as in the 1920s, the Landes producers were content to advantageously export their turpentine to Germany. However, in the 1930s, the cheaper American and Spanish turpentine completely replaced the French one on this market.

To remedy this situation, the Pine Institute undertook in 1934 a collaboration with a new trading company, *Le Camphre Français* (The French Camphor), established by the Ministry of National Defence, which wanted to secure a national camphor supply for the manufacture of explosives.[63] Despite this promising development, the Pine Institute Bulletin did not mention camphor after 1934. The reason for this is unclear, but one plausible hypothesis is that the camphor manufacturing process produced low-value byproducts, known as "residual spirits", which could compete with turpentine spirit as a solvent. While the camphor industry aimed to support the resin industry by using its main product, it would have also created competition in the paint and varnish market. Brus indicated that, in his collaboration with The French Camphor, this problem would be

[63] J.-M. Michel, "Les problèmes du celluloïd", Contribution à l'histoire industrielle des polymères en France, p. 6 (www.societechimiquedefrance.fr/IMG/pdf/a_1_324_100.vfx2_sav.pdf last accessed 18 February 2021).

considered a priority [Brus, 1934]. It is possible that the failure to address this challenge hindered the emergence of the French camphor industry, which was intended to closely work with the resin industry.

In any case, Brus was dissatisfied, and, in 1938, he complained that France was still dependent on importations from Germany and Italy: "In case of a war, we would have to improvise a production plant or buy very expensive — if it is even possible — Japanese or American camphor." [Brus, 1938]

Terpineol, Pinane and Other Turpentine Derivatives

Camphor was not the only compound that could be obtained from pinene that held important promises for the industry in distress. Two other most important derivatives over the period were terpine and terpineol [Desalbres, 1929] (Figure 23). Terpine was used to treat respiratory diseases, but most importantly, it could be transformed into terpineol by dehydration.

Terpineol is a colourless oil used in perfume-making, thanks to its pleasant smell. Its stability with alkali also made it popular among soapmakers, but its most important outlet was the manufacture of synthetic resins, varnishes, and terpinyl ethers [Dupont, 1926, p. 306].

In 1924, the Pine Institute undertook the first studies to facilitate the transformations of turpentine into these different products, but the corresponding industries had to be first convinced of its potential [N. N., 1925]. A new remarkable outlet, however, emerged in the

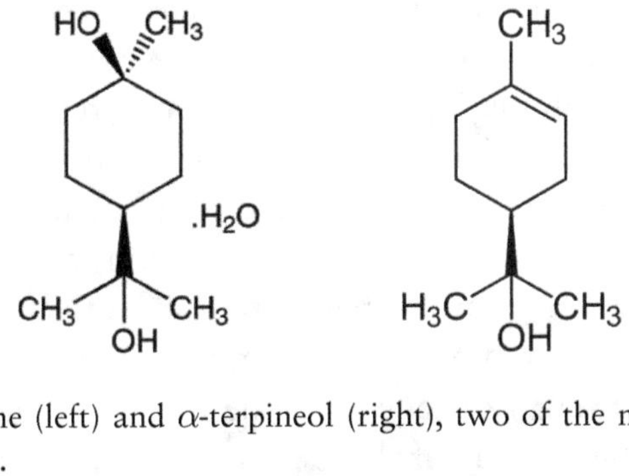

Figure 23: Terpine (left) and α-terpineol (right), two of the most promising turpentine derivatives.

early 1930s when the so-called American pine oil, extracted by the distillation of pine trunks, entered the French market. In fact, the French used the English name "pine oil" to differentiate it from their own *huile de pin*, which corresponded to a range of different products extracted from rosin. The pine trunk distillation industry did not exist in France but developed rapidly in the US. In this industry, the resinous products (turpentine, rosin or others) were obtained by distilling dead pine trunks left after forest clearings. This new American pine oil quickly penetrated many types of activity in France, such as papermaking, textiles and agriculture. Pine oil was notably incorporated into the famous Bordeaux mixture (*bouillie bordelaise* in French), a pesticide widely used in the vineyards near Bordeaux, because of its disinfectant properties and its ability to reduce surface tension while increasing the mixture's adhesiveness [Brus and Barraud, 1937]. This "wetting power" was also potentially interesting for soaps.

Brus and the Pine Institute researchers found the dependence of local wine and soap manufacturers on American pine oil to be unacceptable. In 1933, they undertook a series of studies to examine the American pine oil and discovered that it was composed of 80 per cent terpineol. Brus noted that French terpineol could easily replace its American counterpart in almost all applications. As a result, terpineol quickly began to be used in Bordeaux mixtures through new collaborations between the Pine Institute and local agricultural research laboratories [Dupont and Dubaquié, 1933].

The threat posed by the Americans stimulated other studies in the 1930s on terpineol and similar compounds in order to develop new insecticides and disinfectants from terpinolene, a byproduct of terpine manufacture that "encumbers the factories, because its utilisation in the industry is strongly limited" [Feytaud and Lapparent, 1935, p. 106].

Were there any other turpentine derivatives? In 1935, Brus remarked that the only ones produced in France were terpine and terpineol (after the failure of synthetic camphor) [Brus, 1935]. He enumerated, however, many other potential outlets: cymene and its derivatives (esters used in perfumes), aminocymen (antiknock agent), dipentene, and even isoprene for the manufacture of synthetic rubber. Nevertheless, none of these products appear in the Institute's

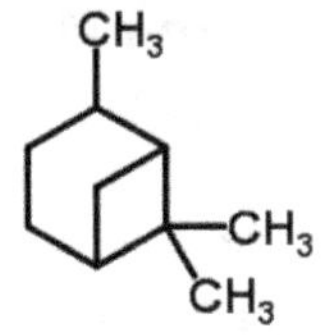

Figure 24: Pinane, a turpentine derivative in which Brus placed much hope.

activity reports. Instead, the Institute was preoccupied the entire 1930s with yet another compound: pinane (Figure 24).

Brus saw great potential in pinane. It was a good solvent that could also be used in fuels, even though its octane rating was only 72, which, according to Brus, was relatively low. However, the extensive research conducted in 1933 did not yield any immediate results in the following years. In 1939, Brus acknowledged the need to continue exploring this avenue, as pinane could still play a significant role in the local economy. Unfortunately, it seems that after six years of thorough investigation, there were no immediate prospects for its application in the region.

Pinane may have primarily attracted scientific interest, and Brus could have emphasised its potential industrial applications to justify the time and resources devoted to its study. It is worth noting that the interwar period fostered research on various types of terpenes, as evidenced by the Nobel Prize in Chemistry awarded in 1939 to the Croatian-Swiss scientist Leopold Ružička, who investigated terpenoids. Although Ružička did not collaborate directly with the Pine Institute, he knew Brus and monitored the Institute's publications on polyterpenes and sesquiterpenes.[64] In other words, terpenes and their derivatives were a fashionable topic, and pinane seems to have held many promises for the Bordeaux researchers who sought to leave their mark in the world of international chemistry.

[64] Gironde Departmental Archives, Pine Institute Archives, file "Correspondance 1938 G. B., R". Letter from L. Ružička to G. Brus, 8 February 1938; Gironde Departmental Archives, Pine Institute Archives, file "Correspondance 1937 G. B., R". Letter from L. Ružička to G. Brus, 4 November 1937.

5. Rosin/Colophony: Discovering the Complexity of Resin Acids

Chemists such as Ružička were not solely interested in turpentine spirit derivatives. In 1928, Ružička attended an international Congress of Wine and Pine (the two most important products of the region) in Bordeaux, where he talked about the constitution of abietic, levo- and dextropimaric acids: the main terpenic acids found in rosin [Ružička, 1929]. In order to obtain turpentine spirit, pine gum is distilled with steam. After distillation, at the bottom of the boiler, there is a solid product called rosin. In fact, rosin constitutes 70 per cent of the resin's weight, which makes it, strictly speaking, the main product of the distillation process.

Rosins were traditionally used in soapmaking, but their value was very low in the nineteenth century before the rise of the modern chemical industry. In the US, before the Civil War, its storage costs were higher than the selling value, and it was treated like an encumbering waste one had to get rid of. The distilleries and workshops dumped rosin into rivers, lakes, or holes dug up in the vicinity [Outland, 2004, p. 159]. In the Landes, rosin was saved thanks to the creativity of local inhabitants. As early as 1822, Etienne Dive, a pharmacist from the small town of Mont-de-Marsan, filed the first French patent on the production of resin oils extracted by the distillation of rosin [Vèzes and Dupont, 1924, p. 462]. Over time, the number of different uses for rosins grew, especially for the manufacture of varnishes. Despite this, it remained a low-value product, considered to be an accessory to turpentine.

Interestingly, once resin chemists undertook the task of studying rosin's composition, their work became immediately relevant from the practical and commercial standpoint.

Directly Applicable Theoretical Studies

In the first years of the twentieth century, the Pine Institute's scientists attempted to assist the resin industry in discovering novel applications for rosin. However, they were confronted with a massive amount of existing literature on the subject. In 1902, Vèzes made a major synthesis of the studies on rosins and wrote that "colophony can be considered the most well understood of the resinous materials: from various bodies that we assemble under the general name of resin, this is the one that the deserves this title the most ... This is the resin par excellence. Some authors call it the resin of the resins" [Vèzes, 1902, p. 1]. And yet, 20 years later, Vèzes and Dupont were far less sanguine about rosin chemistry, considering it much less understood in the light of new studies [Vèzes and Dupont, 1924, p. 400]. Why was the study of rosins problematic? Vèzes and Dupont tried to elucidate the question:

> [T]he obscurity that reigns in this branch [of knowledge] has a number of reasons:
>
> 1. Great instability of the solid elements of the gum, which, under the influence of heat or various agents, isomerise easily.
> 2. Their high level of oxidability.
> 3. The fact that most of these elements are isomers and correspond to a common formula $C_{20}H_{30}O_2$.
> 4. The difficulty that we have, in general, to obtain crystalline derivatives and, as a consequence, well-defined chemical forms.
> 5. When, on the contrary, we obtain derivatives sufficiently stable and well-crystallised ... We find out that we deal, generally, not with defined chemical species but with mixtures of isomorphic isomers that crystallise together in all proportions [Vèzes and Dupont, 1924, p. 400].

Another problem encountered by the chemists was the differentiation between *galipot*, small crystal chunks in the resin gathered from the tree, and rosin, a solid part of the resin after the distillation. Due to the thermal treatment, the composition of resin acids changes, making the exact composition of *galipot* unknown for a long time. Thus, it was necessary to distinguish the turpentine acids (*acides térébenthiniques* in French) present in the *galipot* and the abietic (or colophonic) acids, which are transformed by heat and found in rosin.[65]

All these resin acids are stereo-chemical tricyclic isomers with the same formula ($C_{20}H_{30}O_2$), whose physical and chemical properties are relatively similar. Extracting individual resin acids is a complex and repetitive process requiring multiple consecutive distillations (and crystallisations) to separate an acid of hypothetical properties. Chemists all over the world spent decades dissecting rosins to identify their exact composition. The Pine Institute tried to put together all its prior knowledge and contribute to this difficult task with its own studies. The history of the discovery of different resin acids is a classic history of science, where a collaborative international effort built, over decades, resulted in more and more convincing theories on the exact composition of a complex chemical substance.

Of course, the proportion of these acids varied widely between different tree species, and while the Pine Institute was not oblivious to other types of rosin, maritime pine from the Landes remained at the heart of its studies. However, this diversity was not obvious back in the nineteenth century. Many chemists ignored the botanical differences and believed the same molecules persisted in different species. This approach led to a great deal of terminological confusion, and careful analysis of the results was necessary to uncover the actual substances they worked with [Vèzes and Dupont, 1924, p. 402]. In 1920, one of the Institute's chemists noted: "The first trait that strikes us when we go through this documentation [on the rosin], is the richness of the nomenclature that contrasts with the imprecise meaning of individual terms." [Duffour, 1920]

Between 1923 and 1927, Dupont and his collaborators established a research programme for the Pine Institute to fill the gap in the knowledge on rosins. It was constructed around four large questions:

[65] The name "turpentine acids" is a direct translation of *acides térébenthiniques*. It is not, however, a terminology admitted in the English-speaking literature because in English the term "rosin" is used for both *galipot* and the solid distillation product. For the sake of clarity, I use the French-inspired terminology throughout the study, but the readers should be advised that the difficulties in translation continue to plague resin chemistry and related fields of study until this day.

1. Do we find the same turpentine acids in different types of resins, and in what proportions?
2. What is the relationship between turpentine acids and abietic acids? Does every turpentine acid lead to a single abietic acid?
3. Do different turpentine acids lead to different abietic acids?
4. In what state do we find resin acids in rosins? [Rouin, 1928, p. 122]

This precise strategy rapidly brought interesting results. The Pine Institute confirmed that only the levopimaric acid, a very active resin acid, transformed into the abietic acid under the influence of heat or strong acids, while the dextropimaric acid, another main component of the *galipot* and rosin, according to the researchers in the interwar period, remained unchanged.

This was but the beginning of a much more complicated history. In 1927, the Pine Institute announced the discovery of a new acid, pyroabietic acid, which would have been the third stage of the thermal transformations of rosin (levopimaric to abietic, and then to pyroabietic) [N. N., 1928, p. 50]. Around 1930, it was confirmed that pyroabietic acid was, in reality, a mixture of three different acids: dehydroabietic, dihydroabietic and tetrahydroabietic [Brus, Legendre, Niolle, 1952]. In 1944, René Lombard from Strasbourg isolated a new but very stable acid, which he called dextrosapinic. He believed he had revolutionised the entire rosin chemistry by claiming that there were only three acids in *galipot*: levopimaric, dextropimaric and dextrosapinic [Lombard, 1944, p. 201] (see Table 6 for more details). Unfortunately, his hypothesis was rapidlycontradicted, and the new acid was shown to be a mixture of abietic and neoabietic acids [Lombard, 1954]. A similar story happened in 1936, when a German team announced the discovery of a new proabietic acid (that had similar properties to another acid supposedly discovered before by Dupont: alepabietic), which was later shown to be a mixture of levopimaric, neoabietic, abietic, and isodextropimaric (known today as isopimaric) acids. All these events show that establishing the composition of pine rosin was far from straightforward, and the conjectures were overhauled a few times over the last century.

Table 6: Evolution of the hypotheses on the composition of rosin and *galipot* from *Pinus pinaster* (the majority of them have the simplified molecular formula of $C_{20}H_{30}O_2$)[66]

Source	Acids in *galipot* (before thermal treatment)	Acids in rosin (after thermal treatment)
August Laurent (around 1840)	Pimaric (100%)	Pimaric (100%)
Georges Dupont (around 1924)	Dextropimaric (11%) Levopimaric (22%) Various sapinic acids (67%)	Dextropimaric (10–20%) Abietic (80–90%)
René Lombard (1946)	Levopimaric (35%) Dextropimaric (30%) Dextrosapinic (35%)	Levopimaric (3%) Dextropimaric (30%) Dextrosapinic (30%) Abietic (37%)
Harris (end of the 1940s)	No hypothesis	Levopimaric (1%) Abietic (30–40%) Neoabietic (10–20%) Dehydroabietic (5%) Dihydroabietic (14%) Tetrahydroabietic (14%) Dextropimaric (8%) Isodextropimaric (8%)
Pine Institute (1952)	Levopimaric (30–35%) Abietic (25–30%) Neoabietic (20%) Dextropimaric (7–8%) Isodextropimaric, dehydrabietic and dihydroabietic (12–13%)	Levopimaric (2%) Abietic (50–55%) Neoabietic (18–20%) Dextropimaric and isodextropimaric (10–15%) Others (5%)
Contemporary studies (2008)	Pimaric/previously known as dextropimaric (8%) Sandara-copimaric (2%) Levopimaric and palustris (39%) Isopimaric (12%) Dehydroabietic (4,2%) Abietic (14%) Neoabietic (18%)	Irrelevant category; all depends on the final products desired

[66]Table established with data in: Gironde Departmental Archives, Pine Institute Archives, box “Divers”. G. Dupont, “Constituants cristallisés de la gemme de pin maritime”; Gironde Departmental Archives, Pine Institute Archives, file “Lettres Techniques, 1952–1953, C”. Letter from Pierre Legendre to Etablissement Castel, 3 February 1953; [Panda, 2008, p. 143].

What is the contemporary scientific consensus? To put it simply, under the influence of heat, levopimaric acid transforms into neoabietic acid, which later transforms into abietic acid. There are then three main stages of transformation during a typical resin distillation process. It is, however, necessary to point out that chemists today do not try to analyse the exact composition of solid resin distillation products because the exact composition of rosins varies depending on temperature, pressure, and the duration of heating; scientists concentrate much more on the extraction of specific acids from *galipot* through fractionation. Chemists in the first half of the century were, of course, aware of these various factors. However, they put considerable effort into defining the correct composition of what rosin should be. In other words, their frame of reference was influenced by the practices of the industry that manufactured rosin as a byproduct of the distillation but whose main purpose was to obtain turpentine. The composition of rosins was then dictated, especially in smaller distilleries, by the practices that ensured the production of large quantities of high-quality turpentine. Employing higher temperatures or longer heating durations was instrumental in preventing any loss of undistilled turpentine spirit within the rosin, ensuring an efficient extraction process. However, it was also crucial to strike a balance and avoid overheating to prevent the contamination of turpentine with resin acids. When the modern industry bypassed these limitations, the traditional category of rosin became less relevant for chemists.

This does not mean that rosin, understood as a mixture of different acids, did not have commercially interesting properties on its own. It was precisely the diversity of its components that constituted its main advantage from the industry's point of view. In a certain sense, the definitions elaborated by chemists concerning the correctly distilled rosin had more normative than descriptive value. Ultimately, correctly distilled rosin is the one that sells. How exactly could the variability of the composition of resins be useful for the industry? Rosin had a tendency to crystallise over time [Dupont, 1925, p. 183]. Dissolving huge chunks of crystal was hard and hindered the use of rosins in soaps, paper pulp, paints, and varnishes. It was one of the issues the Pine Institute was regularly consulted about.

The problem was studied in 1925 by Dupont. His research led him to elaborate a series of recommendations directed towards manufacturers. He warned against overheating or not distilling the resin completely, but his results were not conclusive and showed significant variability [Dupont, 1925]. It was the studies on the composition of resins, especially in the 1940 and 1950s that allowed chemists to better understand why some rosins crystallised better than others. In fact, it is precisely the heterogeneity of acids that impedes crystallisation. In other words, a chunk of rosin that was insufficiently heated (and had too much of the levopimaric acid) and an overheated one (with too much of the abietic acid) would crystallise more frequently than the one in which the acids are in balance. By the 1960s, the consensus was that the best rosins were the ones in which different acids (levopimaric, neoabietic and abietic) were in identical proportions. To determine at which point the distillation should stop, the Institute studied the plots of the optical rotation of the mixtures of different acids. It was shown that there was a direct correlation between the value of the optical rotation of the rosin and the diversity of acids it was made of (see Table 6).[67]

All these findings were communicated to local companies on a regular basis. For example, in 1953, a client of one of the biggest Landes companies, Castel, complained about receiving a partly crystallised rosin, prompting them to consult the Pine Institute.[68] Pierre Legendre, the technical director of the Pine Institute, advised Castel to avoid excessive isomerisation of the acids. In another letter, Castel asked Legendre how they could do this, to which Legendre responded that some isomerisation was normal and unavoidable during distillation and suggested controlling the temperature and being mindful that isomerisation continued during cooling. Legendre's response, which spanned several pages, served as an introduction to the chemistry of rosins, indicating that this

---

[67] Gironde Departmental Archives, Pine Institute Archives, file "Lettres Techniques 1966–1967, C". Letter to Coopérative Résineux-Roquefort, 14 January 1967.

[68] Gironde Departmental Archives, Pine Institute Archives, file "Lettres Techniques 1952–1953, C". Letter from P. Legendre to Castel, 5 February 1953.

basic knowledge was not always common even among major local producers.

Legendre created a comprehensive table to summarise the evolution of the knowledge on the composition of rosins. The first and the last line aside, Table 6 is established from Legendre's letter from 1953.

Table 6 serves as a visual representation of the ongoing development and progression of our understanding of the composition of *Pinus pinaster* rosins over the past century. The increasing complexity of the acids listed in the table reflects the discovery that many previously identified acids were, in fact, mixtures of other acids. Furthermore, the definitions of certain acids were revised over time as a result of new laboratory techniques and continued research efforts. A similar table could be established for every other species giving commercial resins, although few others were given that much attention (*Pinus palustris* and *Pinus sylvestris* are notable examples).

Taming Rosins: Polymerisation, Hydrogenation, and Dismutation

Studies on the proportion of various acids in rosin were of theoretical interest, and yet they had direct applications. This is also the case of another big line of research concerning rosin: the study of the exact structure of the acids in question. Figure 25 shows the contemporary

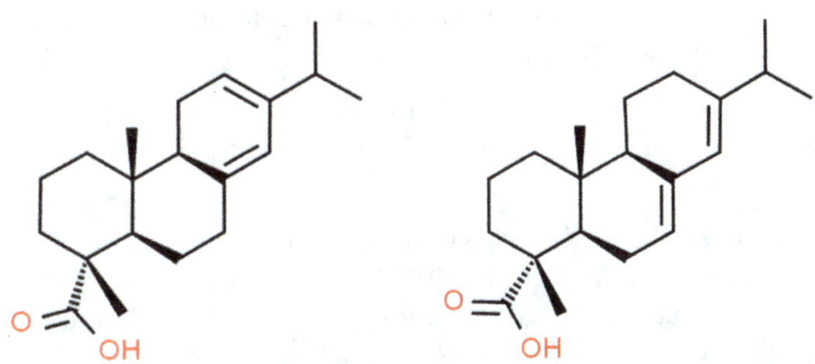

Figure 25: Levopimaric acid (left) and abietic acid (right). The unsaturated bonds are denoted with a double line.

structures of the two most significant acids in rosin: levopimaric and abietic.

Without delving into too much detail, it is immediately evident that the only difference between them is the position of two unsaturated bonds, which make these acids very reactive and prone to oxidisation [Panda, 2008, p. 186]. Exposed to air, rosins darken, and their value decreases significantly, as colour is the most important factor in the final price [Dupont, 1927b]. The oxidation process, similar to crystallisation, directly impeded the use of rosin in soap and paper-making.[69] The Pine Institute took on the challenge and focused on the methods of saturation of these problematic chemical bonds. As in the previous case, this topic was not studied exclusively in the Institute, but the Institute's role in gathering different results and transferring them to the practices of the local industry was unquestionable.[70] Three techniques were employed to "stabilise" the resin acids, primarily levopimaric acid: hydrogenation, dismutation/disproportionation and polymerisation. In the following section, each method is described in detail.

The oldest technique used to saturate bonds in resin acids was hydrogenation. It entailed adding hydrogen to form dihydroabietic and tetrahydroabietic acids, depending on the number of saturated bonds (Figure 26).

In 1891, one patent was filed for the "reduction of resins" (meaning hydrogenation in this context) but without giving technical details on the methods used.[71] In the interwar period, this technique was studied by some of the biggest resin-producing companies in the world, such as Hercules Powder Company in the US. The Pine Institute began to investigate the topic relatively late, only in 1937, but pursued this line of research for many years.

---

[69] Gironde Departmental Archives, Pine Institute Archives, file "Lettres Techniques 1954–1955, J". Letter to Joly, 4 October 1955.

[70] Gironde Departmental Archives, Pine Institute Archives, file "Travaux IP: Brus, Legendre". Summary of studies from Legendre's laboratory in 1951.

[71] Gironde Departmental Archives, Pine Institute Archives, file "Colophane hydrogénée", box "Acides Résiniques". G. Brus, P. Legendre, "Sur la Colophane Hydrogénée", Congrès International des Peintures, October 1947.

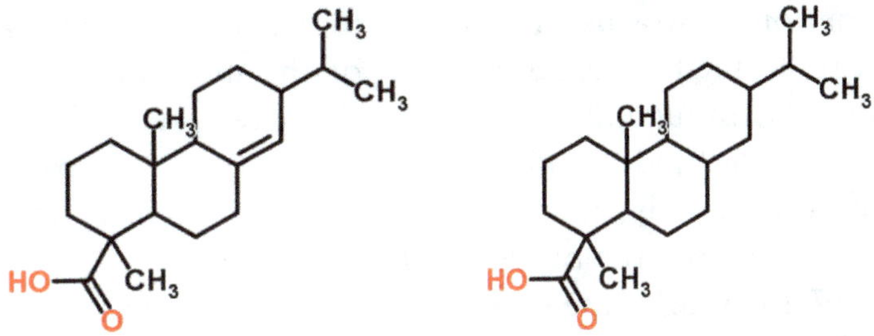

Figure 26: Hydrogenated resinic acids: dihydroabietic acid (left) and tetrahydroabietic acid (right).

In total, the Pine Institute's archives indicate that between 1916 and 1954, only seventeen patents on rosin hydrogenation were filed worldwide: one in the United Kingdom, three in Germany, five in France (but only one by the Institute's researchers), and eight in the US. The stabilised rosin Staybelite, patented by Hercules Powder in 1928, is still being sold today. The Pine Institute's patent on the "Hydrogenation of resin acids, of the products that contain them, and are derived from them" was filed in 1945 and granted in 1947 [Brus and Legendre, 1947]. Its originality, as compared to other similar patents, is that it allowed for "hydrogenating at a lower temperature and avoiding the decarboxylation of resin acids and the crystallisation of the dihydro-dextropimaric acid" [Brus and Legendre, 1947, p. 1].

The hydrogenated rosin, which had a significant proportion of hydrogenated abietic acids, shared similar properties with regular rosin in terms of colour, acid value, hardness, and dropping point.[72] It was, however, much more stable and did not oxidise. The potential of this technology was huge for the local industry striving to enhance the quality of its products. In order to fully exploit the patent, a trading company was established in 1943: the *Société d'études pour l'hydrogénation des colophanes* (Society for Studies on the

[72] Gironde Departmental Archives, Pine Institute Archives, file "Colophane hydrogénée", box "Acides Résiniques". G. Brus, P. Legendre, "Sur la Colophane Hydrogénée", Congrès International des Peintures, October 1947.

Hydrogenation of Rosins). It brought together the members of the Pine Institute and the representatives of the most important local companies.[73] Although there are no available records of the Society's exact activities, it funded and facilitated the research on rosin hydrogenation. Additionally, it is clear that the patent filed by Brus in 1945 was considered the property of the Society.

In 1947, the year of the patent's publication, the French chemical giant Saint-Gobain expressed an interest in buying it. The Institute's director, Brus, was in favour of the transaction and believed that Saint-Gobain had the capability to conduct semi-industrial tests necessary to identify potential problems that could arise during hydrogenation in chemical plants. And yet, other members of the Society for Studies on the Hydrogenation of Rosins, notably Jacques Boulin and his engineer René Escourrou, from the most important local company *Dérivés Résiniques et Terpéniques* (DRT — Resin and Turpentine Derivatives), insisted on the patent's value for the region and believed that semi-industrial tests should be conducted locally. Brus was against this and explained that the Landes industry did not have enough experience in constructing large-scale industrial plants.[74] Boulin suggested that the Saint-Gobain chemists conduct the semi-industrial tests without being granted the patent. The rest of this story remains unknown, but the correspondence between DRT and Brus suggests that the DRT decided to work on this topic independently from its other partners.[75]

While Brus battled for the hydrogenation of rosins, the former director of the Pine Institute, Dupont, argued that the hydrogenation of rosins was too expensive and saw a greater potential in another

[73] Gironde Departmental Archives, Pine Institute Archives, file "Colophane hydrogénée", box "Acides Résiniques". Document "Société d'étude pour l'hydrogénation des colophanes".

[74] Gironde Departmental Archives, Pine Institute Archives, file "Colophane Hydrogénée", box "Acides Résiniques". "Rapport de la réunion du 6 janvier 1947 au cabinet du Doyen Brus de la Société d'étude pour l'Hydrogénation des Colophanes".

[75] Gironde Departmental Archives, Pine Institute Archives, file "Colophane Hydrogénée", box "Acides Résiniques". Letter from Désalbres to Brus, 6 February 1948.

technology: disproportionation.[76] This technique can be traced back to 1928, in a serendipitous discovery by Dupont and Jean Dubourg, head of the "Rosins and derivatives" section of the Pine Institute, in 1928. "We had a chance to find, in the bottles of strong resin oil, abandoned 20 years before, a perfectly crystallised isomorphic isomer of the abietic acid, whose optical rotation was clearly dextrorotatory" [Dupont and Dubourg, 1928]. Initially called the pyroabietic acid, the Institute's research led to the conclusion that it was, in reality, a mixture of three acids: dihydroabietic, tetrahydroabietic, and dehydroabietic [N. N., 1928, p. 50]. Dupont continued his work on this last acid after he departed from the Pine Institute in 1933, and during his tenure at the École normale supérieure in the 1950s, he filed several patents for the manufacture of dehydroabietic acid and its derivatives (Figure 27).

"The process involves direct hydrogenation of resin esters with sulphur or bromine or by heating in the presence of the dehydrogenation catalyst," explained one of Dupont's patents. The result was a very stable aromatic ring. According to Dupont, this acid would open a range of previously closed market outlets, in particular in the manufacture of plasticisers.[77] At least one company in Aquitaine

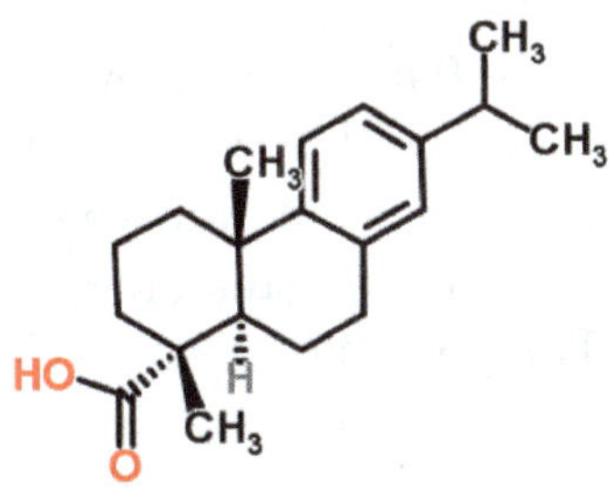

Figure 27: Disproportionated rosin (dehydroabietic acid).

[76] Gironde Departmental Archives, Pine Institute Archives, box "Divers". G. Dupont, "Sur l'acide déhydroabiétique et quelques dérivés", extract from *Bulletin de la Société chimique*, n. 1, 1958.

[77] Gironde Departmental Archives, Pine Institute Archives, box "Divers". G. Dupont, "Sur l'acide déhydroabiétique et quelques dérivés", extract from *Bulletin de la Société chimique*, n. 1, 1958.

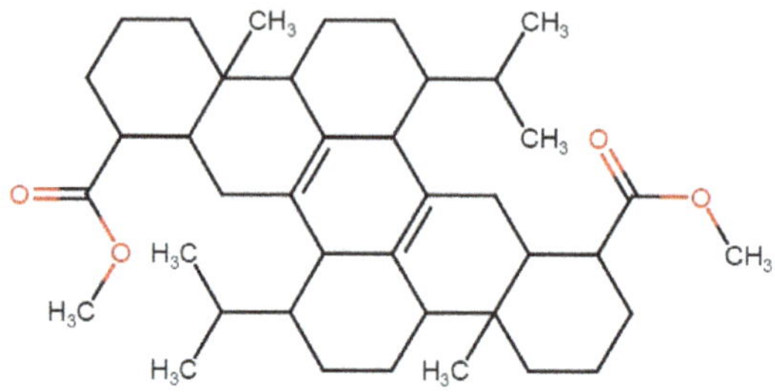

Figure 28: Di-abietic acid.

delivered to the market dismuted rosins with an important fraction of dehydroabietic acid in the 1950s and 1960s.[78]

In the 1950s, while Dupont was patenting his discoveries, Brus and his team were exploring another method of saturating resin acids — polymerisation. In 1953, they successfully synthesised di-abietic acid [Brus *et al.*, 1953] (Figure 28). Brus was confident that polymerised rosins would be in high demand from soap producers and the paint and varnish industry and could also be used to produce new derivatives such as salts, esters, and condensed resinous products.[79] This invention marked a slow shift towards the study of polymers that started becoming the most important object of the Institute's studies after the fall of the resin industry in the 1970s.

And yet, despite promising laboratory results, the polymerisation of rosins remained a challenge on an industrial level in the 1950s and 1960s.[80] The difficulty lay in controlling the proportion of dimers in a given rosin sample. The polymerisation level had to be studied by observing the softening point, which was higher if the rosin was rich in di-abietic acid. However, the softening point was affected by a

[78]Gironde Departmental Archives, Pine Institute Archives, file "Correspondance 1958, C". Letter from Cellulose du Pin, 25 November 1958.

[79]Gironde Departmental Archives, Pine Institute Archives, file "Travaux IP: Brus, Legendre". G. Brus, "Travaux récents effectués à l'Institut du Pin, Conférence du Syndicat des Sylviculteurs avril 1950".

[80]Gironde Departmental Archives, Pine Institute Archives, file "Lettres Techniques 1958–1959, C". Letter from Institut du Pin to La Cellophane, 29 June 1959.

range of different factors, including pollution with non-acidic compounds. The Institute had to use UV absorption to study the proportion of abietic acid and impurities in a given sample. By combining these two methods (softening point and UV absorption), it was possible to approximately measure the dimer content. Unfortunately, these were complex laboratory tests, and the companies purchasing polymerised rosins often faced challenges in verifying the quality of the product on-site.

Resinates (Gum Rosin Salts)

Rosins, regardless of whether they are saturated or not, are not directly usable in many industries; for example, while rosin can improve the consistency of paints, "it is gluey in warm hands and sticks to fingers, a defect found in layers of varnish containing some proportion of rosin. Moreover, it is not water repellent and, by its constitution, it is too acidic, which prevents its use in the manufacture of paints destined for cement surfaces and prevents the addition of basic pigments [such as zinc oxide, white lead, and minium]," explained an article published in 1925 in the Pine Institute Bulletin [Blain, 1925]. It was necessary to harden rosins and decrease their acidity by transforming resin acids into salts known as resinates.

In the early 1920s, it was possible to manufacture the so-called molten resinates by mixing the rosin heated up to 200 °C with an oxide (*e.g.*, MnO_2, $Mn(OH)_2$, PbO, CaO, ZnO) [Dubourg, 1927]. This method was frequently employed to obtain zinc or lime resinates; however, the resulting salts were still too acidic because of the proportion of free acids. In 1924, the Pine Institute patented another technique for creating the so-called precipitated rosins [N. N., 1924c]. First, the rosin was treated with a sodium carbonate solution to obtain sodium resinate soluble in water. Then, the soap reacted with a heavy metal salt in order to resinate the metal in question [Dubourg, 1927, p. 122]. These resinates were significantly less acidic than the molten ones.

There were two primary applications for resinates, which enjoyed industrial synergies in the Landes. First of all, alkaline resinates (resin soaps that were very similar to normal fatty acid soaps) were used as domestic soaps, but more importantly, as a component of paper sizing, a substance that reduces the paper's tendency to absorb liquid. The sizing with resin soaps is carried out to cover the paper with a molecular layer impermeable to water. Not only does this layer give the paper a certain resistance to humidity, the former also prevents the latter from absorbing the ink in the printing process too rapidly. The paper industry was (and still is in many places) the most important outlet for rosins. In fact, its introduction in the Landes of Gascony in the 1920s was not straightforward since many believed that pine wood was not good for papermaking, but access to cheap sizing was one of the arguments in favour of its development [N. N., 1924c]. The second primary application of resinates was copper resinate, which was utilised as a wetting and fixative agent in agriculture, particularly in the treatment of fruit trees during winter. The Pine Institute worked closely with the Phytopathologic Station of Bordeaux over several years to enhance this technique [Dupont and Dubaquié, 1933; Feytaud and Lapperent, 1935].

Metallic resinates aside, in 1924, the Pine Institute also explored glycerin resinates, also called resin ethers [Lafon, 1924], which were "molten resinates in which the neutralising agent is not a metal, but an alcohol glycerine. The etherification happens at around 290 °C in the presence of a small amount of metallic oxides (CaO, ZnO)" [Dubourg, 1927, p. 121]. Their manufacture was cheaper than metallic resinates, and with the help of the Pine Institute, a workshop manufacturing "etherified rosins" was opened in the late 1920s in the small town of Labouheyre (Landes) to supply the paint and varnish industry. It is important to underline, however, that these resin ethers could not replace metallic resinates in every application because they were too soft.

Rosins and Road Coatings

So far, three major rosin-related research topics were explored: rosin composition, rosin saturation, and resinates. All of these were driven by both scientific interest and practical needs of the industry. The last topic, however, is strictly practical and pertains to the use of rosins in road coatings. In the 1920s and 1930s, the development of road coatings was on the national political agenda in France due to the rapidly growing number of cars [Trouis, 1927]. During this period, road coatings were made of emulsions that contained "in general, 50 per cent water and 50 per cent useful materials: bitumen or tar mixed with an emulsifier. In contact with the ground, the emulsion breaks and the layer of bitumen or tar separated this way adheres to the materials that the road is made of," explained Brus and Barraud in 1935 [p. 13].

The researchers at the Pine Institute became interested in exploring the use of rosins in road coatings as early as 1921 when Dupont was studying the possibility of replacing some proportion of tar with dark rosins that were inexpensive and had limited practical applications. Dupont advocated the use of a rosin-based solution called Pulvorite in road coatings during his directorship. However, because of the rapid rise of rosin prices in the early 1920s, the project was abandoned. By the early 1930s, the situation changed, and rosin prices reached an all-time low. This prompted Brus and Barraud to investigate how rosins could be used in road coatings to help the resin producers recover from the crisis. In 1935, they formulated several strategies to encourage the industry to use rosins in road coatings, taking into account technical, economical and political factors. The Pine Institute researchers saw themselves responsible for the fate of the local industry and wanted to help by all possible means, lobbying included [Brus and Barraud, 1935, p. 13].

Notably, Brus developed a new process in 1934 for replacing 10 per cent of the overall weight of the road coating with dark rosin at the expense of tar or bitumen. The same year, local road administrations in different departments of Aquitaine agreed to conduct large-scale experiments on the quality and durability of road coatings based on rosin. By 1935, Brus was optimistic that the tests showed that the coatings could absorb up to 20 per cent of the rosin without negatively affecting their quality. However, he underlined the need for more tests [Brus and Barraud, 1935, p. 14].

And yet, the use of rosins in road coating also encountered criticism. Some argued that they had a low softening point and friction resulted in acid dust harmful to engines. The solution would be to use resinates instead of rosins [Ortal *et al.*, 1935]. This, however, would have increased the overall costs. Additionally, some studies found that the addition of resinous products to tars is not favourable because they tend to oxidise, decreasing the quality of the coatings over the years [Brus, 1936, p. 161]. While preventing oxidation through hydrogenation was an option, the technology was not mature enough in the late 1930s and would have increased costs as well. Although Brus argued that more studies had to be conducted, these criticisms must have contained a grain of truth because he later changed the tone of his communications and started insisting on the fact that resin was produced on the French soil, unlike bitumen, which was imported [Brus, 1936, p. 161]. Brus's arguments were no longer about the quality or saving cost but about national interests.

Despite this attempt, by December 1938, the case for rosins was definitely lost, and the Institute reoriented its studies. Brus, admitting that rosins were too expensive to replace the tar in the coatings, proposed a new approach. Instead of replacing tars or bitumen entirely, Brus suggested using sodium and potassium resinates as emulsifiers, in a quantity that would correspond to around 1–2 per cent of the overall weight of the coating [Brus, 1938, p. 259]. While this alternative approach was interesting, it would have used considerably less rosin than the previous proposal, which aimed for up to 20 per cent of the coating's weight.

In 1939, research on the new application of rosin continued to focus on three areas: improving adherence to the road, enhancing gravel fixation, and ensuring the homogeneity of the coating [N. N., 1939]. In 1940, due to the shortage of oil-based materials, rosin was again a candidate for road coating on a larger scale, but this was the last time the issue was mentioned in the Pine Institute Archives [Brus, 1940, p. 8]. It appears, however, that some companies continued to manufacture rosins adapted for road coating after World War II, at least until the 1950s.[81]

[81] Gironde Departmental Archives, Pine Institute Archives, file "Lettres Techniques 1956–1957, P". Letter to Ingénieur en chef des ponts et chaussées en Alger, 23 November 1957.

Conclusions: Resin Chemistry — A Forgotten Science of Sustainability

In this chapter, the validity of the "linear model of innovation" has been questioned. This model suggests that applied research follows from basic or fundamental research.[82] However, in the case of French resin chemistry during the interwar period, fundamental research questions were closely intertwined with practical concerns and market considerations. If the local industry did not have to fight against substitutes of turpentine spirit, its properties and composition would not have been so meticulously and thoroughly studied. It was because the industry struggled with the crystallisation of rosins that the local scientists analysed, step by step, the thermal changes in resin acids, discovering a host of new acids in the process. It was through the work on different ways of preventing the oxidation of rosins that the location of unsaturated bonds in resin acids could be studied.

Although the reversal of the relation between basic and applied research, as in the case of resin chemistry, is far from unique in the history of science (one could argue that this is the dominant model), what is surprising here is that while all this research was driven by the needs of the industry, it was not necessarily steered by the companies themselves. Resin chemistry stood out as something one could call "university-driven industrial science". What does it mean? The Pine Institute chemists, attached formally to the local faculty of science, viewed themselves as mentors, guides and leaders of the entire economic sector and tried to shape it with their own ideas. It was not

[82] The linear model of innovation is one of the classic topics in history of science and technology. According to the linear model, curiosity-driven disinterested basic research is the foundation for applied sciences and marketable technologies. In this view, the technologies are merely a translation of "pure science" into practice. This is the romantic view of science as a driving force of human progress. The reality is far more complex as what is often referred to as "basic research" is often closely intertwined with practical concerns. Or to put it differently, the differentiation between pure and applied science is rarely clear-cut. See more in [Edgerton, 2004] and [Godin, 2006].

the local companies that came to the Pine Institute with the idea that camphor or road coatings could be potential market outlets for resinous products. On the contrary, it was the scientists themselves who sought to find new ways to reinvent the local economic landscape. This was particularly true for the new method for resin fractionation, which was conceived as a top-down revolution in the way resins were treated. It was strictly applied research, but it did not originate from the industry, nor was it initially desired by the industry, for that matter. Instead, it was designed and promoted by university chemists who were deeply passionate about the potential of resins. The case of Dupont, who, for many years after having left the Pine Institute, still pondered rosin's dismutation methods and hoped to make them work, is an exemplary case.

The question remains about the disciplinary identity of the French practitioners of resin chemistry. It is another interesting element in this puzzle. Chemistry is usually considered a laboratory science par excellence; the word "laboratory" was associated with chemistry before all other sciences. However, some disciplines generate knowledge outside the laboratory. Geology [Porter, 1978], glaciology [Hevly, 1996], and, of course, biology were for long above all "field sciences" [Kohler, 2002]. However, the tension between the laboratory and the field was also at the heart of resin chemistry. The immediate laboratory results of chemists specialised in resin chemistry concerned pines from a very specific region, and even then, they varied immensely when individual trees were compared.

In order to construct any meaningful study aspiring to some level of universality or at least comparability, resin experts had to incorporate the broader context of the environment in the largest possible sense (natural and economic). They regularly worked outside their laboratories in the forest itself. They had to take into account and understand forest workers' practices, the biology of the pines, the impact of climatic and meteorological conditions, distillation techniques, but also the expectations of the consuming industries. Not to mention that the Pine Institute scientists had to navigate the world of trade standards to establish the "correct" products and were actively involved in almost political debates on the place of resins in the national economy. Their vision was profoundly interdisciplinary, multilayered and integrative.

How can the particular character of resin chemistry be explained? It seems that long before the name was invented, resin chemistry was what we could call today a "sustainable" science — a scientific discipline whose aim was to preserve a specific ecosystem while, at the same time, constantly looking for new ways to make it relevant from the economic standpoint. Practitioners of resin chemistry took a holistic approach, considering the local socio-environmental system in its entirety. They did not want to isolate it from external influences or cut it off from the gears of the global economy. On the contrary, they wanted to preserve the local forest, save the jobs of the local inhabitants, and, most importantly, exploit a versatile renewable raw material by showing its potential through scientific research. Resin chemists were not in service of an industry that dreamed of global expansion. Rather, they were striving to create a sustainable, self-sufficient, and resilient socioeconomic and environmental system, an aspiration that is surprisingly modern. They were no longer just chemists but practitioners of a broader resin science.

The two following chapters show how the evolution of the economic and industrial context slowly put these aspirations in question and how the French resin chemistry disappeared during the decades following World War II.

Chapter 3

LOOKING FOR A NEW IDENTITY: THE PINE INSTITUTE DURING AND AFTER WORLD WAR II

> "The chemistry of resinous products has to follow the evolution of the chemical industry, and more specifically, the one of the petroleum products."
> [N. N., 1949]

How did World War II and the reconstruction that followed influence the Pine Institute? How did resin chemistry change through these events? Both were marked by a deep shift in the identity of the Pine Institute's chemists. The introductory quote is taken from the journal *Bois et Résineux*, which bears witness to its endorsement of the rapprochement with the petroleum industry. The idea was that the Institute would reshape resin chemistry in the image of petrochemistry, and Georges Brus hoped to lead it into the new era by collaborating with one of the early petrochemical giants, Jupiter. Yet the Institute's position was more complex, and over the years, its attitude towards the local resin industry shifted multiple times. It is clear, however, that the Institute tried to redefine its identity in a rapidly evolving economic and scientific landscape of the postwar era.

1. Nazi Occupation and Union Corporative des Résineux (U.C.R.)

In the years preceding World War II, the resin industry in Aquitaine was regularly perturbed by conflicts between different stakeholders: forest workers, forest owners, and companies and workshops of various sizes. This fragmentation destabilised the market and made it difficult to form a united front against foreign products and oil-based substitutes. In order to overcome the differences between stakeholders and stabilise the prices, a new organisation was formed in May 1939: *Union Corporative des Résineux* (U.C.R.) [N. N., 1939b]. The U.C.R. grouped 104 plants, factories and workshops producing resin products in France (and in Algeria), as well as the forest owners' associations and forest workers' labour unions. Defining the U.C.R. is not straightforward. For its members, it was "a new organism, based on the corporative principle harmonised with current needs: it is a general interest group that accepts the arbitration of a [directing] committee" [N. N., 1940]. The U.C.R. essentially operated as a trading trust, with its primary objective being to control resin prices and determine the share for all parties involved. The governing committee of the U.C.R. consisted of representatives from various stakeholders: four forest owners, four forest workers, four manufacturers, three distributors, and one broker ("courtier") [N. N., 1939c]. This organisation held significant influence in its interactions with clients, given its ability to regulate prices, and played a crucial role in maintaining harmony among its diverse stakeholders by ensuring the fair allocation of income generated by the resin trade. As a result, the U.C.R. wielded considerable economic and political influence.

The Nazi occupation only reinforced its position. In September 1940, Philippe Pétain famously announced that "France will become again what she should never have ceased to be: an essentially agricultural nation". The chief of the Vichy France[83] took a personal

[83] Vichy France was a puppet state headed by Marshal Philippe Pétain during World War II, which was located in the south-central region of modern-day France.

interest in the resin industry. In December 1940, he signed a decree on the organisation of the resin industry, attributing official powers to the U.C.R. and naming a certain Charles de Marcy its director [N. N., 1940b]. The following year, Pétain named Charles Prat (former C.G.T.[84] activist) the general secretary of *Fédération des gemmeurs et métayers du Sud-Ouest* (Federation of the Forest Workers of South West), the largest labour union federating forest workers. Prat became the U.C.R.'s director after the war.

In 1941, de Marcy wrote in *Bois et Résineux*: "I was deeply impressed by the clarity of mind of Marshal Pétain, by his goodness, by the sense of security and wisdom emanating from his person." [Marcy, 1941] De Marcy explained that the U.C.R.'s organisation was fully in line with Pétain's philosophy of "corporatism". Corporatism (as conceived by the Vichy regime) was meant to replace the concept of class struggle and was founded on the idea that the representatives of different professions in a given sector (workers, technicians, chiefs) could unite and form a common front to defend their interests. This was exactly what the U.C.R. aspired to be. During the war, it had a unique monopoly on the entire trade of resinous products. It notably collected resin from producers, including the requisitions for the German army, and anyone who wanted to buy some had to go through the U.C.R.'s selling centre, provided they had special ration tickets.[85] One could say that the U.C.R. lived its golden age under the Nazi Occupation.

However, resin trade aside, the U.C.R. had one more important mission: to fund scientific research of the Pine Institute. The strong position of the U.C.R. during the Nazi occupation helped to guarantee stable income for the Pine Institute and strengthen its place in the industry. Brus was considered an expert by the Vichy administration and participated in at least one inter-ministerial meeting on the

[84] Confédération Générale du Travail (CGT), General Confederation of Labour in English. One of the largest French labour unions long affiliated to the French communist party.

[85] Website of Conseil général des Landes: www.histoiresocialedeslandes.fr/p5_occupation_01.asp, last accessed 18 February 2023.

installation of the resin factory in Provence (Bordeaux was not in Vichy France but in the occupied zone while the Landes Forest was divided between the two).[86] More interestingly, the U.C.R. issued an official recommendation calling the Pine Institute "the only official laboratory of the Corporation" and indicating that all expertise requests concerning resinous products should be directed to the Institute [N. N., 1942]. In fact, the Institute was presented as the U.C.R.'s internal laboratory. In return, the Pine Institute published in *Bois et Résineux* — which from 1940 became the official press organ of the U.C.R. — its activity reports explain how it contributed to the progress of the resin industry. The Institute also promoted its technical solutions via the U.C.R. by issuing recommendations to local producers, as the U.C.R. saw itself as responsible for ensuring the quality of resinous products.

The rapprochement with the U.C.R. aside, what was the overall impact of the occupation on the Pine Institute? On the one hand, half of all resinous products were requisitioned for the needs of the German industry, weakening and impoverishing the local industrial network in Aquitaine [N. N., 1943]. On the other hand, the Institute was regularly solicited for help by the Italian navy and the German army for small chemical analyses unrelated to resins, for which it was remunerated.[87] Overall, the financial situation did not seem to have deteriorated. Not only that, it was precisely at the beginning of the war that the Pine Institute established long-term collaboration with the young *Centre National de la Recherche Scientifique* (CNRS — National Scientific Research Centre), a leading state research agency that profoundly transformed the French scientific

[86] Gironde Departmental Archives, Pine Institute Archives, file "Correspondance 1942, M". Letter from de Marcy to G. Brus, 24 September 1942.

[87] Gironde Departmental Archives, Pine Institute Archives, file "Correspondance 1941, I". Letter from 23 December 1941; Gironde Departmental Archives, Pine Institute Archives, file "Correspondance 1942, F". Analyses throughout 1942 in favour of Forges et chantiers de la Gironde; Gironde Departmental Archives, Pine Institute Archives, file "Correspondance, 1942, L". Letter from 18 August 1942; Gironde Departmental Archives, Pine Institute Archives, file "Correspondance 1943, O". Correspondence of Omnium Métallurgique des ferrailles & métaux in 1943.

Table 7: Pine Institute's staff in 1941[88]

Staff employed by the Faculty of Sciences	Brus (director), Faucounau (vice director), Arrivaut (honorary assistant professor)
CNRS staff	Barrière, Chollet, Joanne
Staff employed exclusively by the Institute	Legendre, Guillot (chemical engineers) Granier, Fournier, Frossars (chemical engineers assistants) Dupuch (secretary) Faydit (shorthand typist) Petitjean, Fouché (lab assistants) Cassan (cleaning staff)

landscape after the war. In 1941, the Institute's staff was more numerous than ever (Table 7), counting sixteen employees, with three of them remunerated by the CNRS.

The war had an additional impact as it brought attention to research areas that were previously overlooked. One such area was the study of wood hydrolysis for the production of methanol fuel, which gained renewed interest during the fuel crisis. Although research on this subject was conducted throughout the 1920s, it experienced a resurgence after 1939 when the French government decided to invest in the first wood hydrolysis plant in the region, following the method developed by the Pine Institute. This resulted in alternative wood fuels becoming a significant focus of the Institute's research program during the early 1940s. However, there was another closely related issue that had an even more substantial influence on the Institute's agenda: rosin cracking and lubricating oils.

2. Petrochemical Romance: Rosin Cracking, Resin Oils, and the *Société des pétroles Jupiter*

To understand why the Pine Institute and the petroleum industry drew closer, we need to delve into the early nineteenth century.

[88] Gironde Departmental Archives, Pine Institute Archives, box "Divers". G. Brus, "Le Fonctionnement et les Travaux de l'Institut du Pin en 1941", 1943.

In fact, in 1822, Etienne Dive, a pharmacist from a small Landes town, Mont-de-Marsan, filed a patent on a technique for the manufacture of resin oil by rosin distillation. This resin oil was a mixture of heavy hydrocarbons, especially of the so-called terebenthyl ($C_{20}H_{30}$). It was commonly used in the first half of the twentieth century as a lubricant, sizing oil, disinfectant, and insecticide. It also had siccative properties, meaning it could replace linseed oil in varnishes [Barraud, 1932]. Georges Dupont was himself interested in its potential and dedicated much of his work in the 1920s to developing a better understanding of lubrication theory, with the aim of producing higher quality oils [Vèzes and Dupont, 1924, p. 547].

This resin oil remained, nevertheless, a poorly defined substance, and its production methods were far from standardised. In 1935, Brus complained that an important electric cable manufacturer who had used no less than 200 tons of resin oil every year had to switch to petroleum-based products because the Landes producers failed to deliver a product with consistent properties [Brus, 1935, p. 196]. Brus believed that rosin oils had the potential to compete with petroleum derivatives in many markets, but in order to do so, further research on the process of rosin cracking was needed to produce better-defined products.

Cracking was a technique known from the outset of the petroleum industry (it involves breaking long organic molecules into simpler ones). In fact, it had been used since the 1850s to produce kerosene, and many patents improving this technique had been filed in the following decades. However, a significant breakthrough occurred in 1927, when Eugène Houdry, a French chemist who moved to the United States (US), patented the catalytic cracking of petroleum feedstocks and successfully applied it on an industrial level in 1931. In the 1930s, cracking became a fashionable topic, and it is not surprising that the concept also excited Landes scientists.

In 1938, J. Brodschi, a doctoral researcher at the Pine Institute, defended his dissertation *Contribution à l'étude du cracking de la colophane* (Contribution to the study of cracking of rosin) [Brodschi, 1938]. He emphasised in the dissertation's introduction that cracking was nothing new in the region since rosins had been subjected to "pyrogenation" or "fractional distillation" for decades. The shift to "cracking" terminology was to highlight the parallels between resin chemistry and petrochemistry. Brodschi explained that the light cracking of rosin made the abietic acid open one of its side chains and transformed it into "octohydroretene". Further cracking in higher temperatures could produce "benzene, cyclohexene, naphtalene and hydro-naphtalene" chains.[89] What matters is that it was demonstrated that rosin could be a source of a range of new byproducts, depending on the catalyst or method used, including lighter spirits potentially usable as fuel.[90] The significance of this discovery lies in the fact that it opened up new possibilities for the use of resin as a substitute for petroleum. Brus expected the technology to profoundly transform the local economic landscape, and he therefore reached out to the largest French petrochemical company, Jupiter (known from 1948 as Shell Française), to explore its potential.[91]

The first practical works on rosin cracking took place in Jupiter's refinery in Pauillac in Gironde in 1941. Nicolas Orlowsky, Jupiter's leading chemist on the project, noted that, after the first catalytic distillation, no one in the refinery had the slightest idea about the composition of the obtained oil. This illustrates how little the petroleum industry knew about resinous products at the time. In fact, Jupiter's internal reference report on lubricating oils included sunflower oil, sesame oil, linseed oil, and even cacao butter as a potential raw material but no resins. Even the experts in the field ignored this traditional use in the Landes region. Despite these obstacles, Jupiter

[89] Gironde Departmental Archives, Pine Institute Archives, box "Shell", notepad 1. Letter to Comité d'Organisation des Carburants & Lubrifiants de remplacement, 30 September 1943.

[90] Gironde Departmental Archives, Pine Institute Archives, file "Correspondance 1938 G.B., G". Letter from G. Brus to dean Godchot, 22 June 1938.

[91] Gironde Departmental Archives, Pine Institute Archives, box "Shell". N. Orlowsky, Société Anonyme des Pétroles Jupiter, Raffinerie de Pauillac: Recherches et Travaux sur la Fabrication des Huiles de Colophane, no date, p. 3.

decided to pursue this line of research, especially given the difficult economic context of the occupation.[92]

The first semi-industrial cracking column, based on Brodschi's design and improved by Brus, was constructed on 15 February 1942. However, due to supply difficulties, the installation had to rely on charcoal for heating, leading to unsatisfying results, especially concerning heavier fractions. The situation became more and more urgent. Other companies, notably *Les Consommateurs du Pétrole* (Consumers of Petroleum), started to experiment with resin oils as well. This is due to the fact that the German authorities controlled the entire resin market, and rosin allocations were attributed in priority to companies trying to transform it into various petroleum substitutes. If Jupiter wanted to obtain more rosin, it had to show promising results. Without them, it would not obtain its rosin share and lose its advantage over competitors, and probably the costs of the installations as well.

Orlowsky explained that in February 1942, Jupiter engineers still lacked the knowledge of how to produce commercial resin oil, but after a few tests, a catalyst was selected (zinc chloride ($ZnCl_2$)) that gave satisfying results. By the end of the month, an official visit took place in Pauillac to evaluate its productive capacities and decide on the attribution of rosin. Brus was among the evaluators, which is not surprising given his expertise on the matter, but his connection to Jupiter raises concerns about the impartiality of his evaluations. The same year, when other companies approached the Institute to exploit its patents, despite long negotiations, Brus refused, and it was Jupiter that was granted an exclusivity contract. Whatever the real situation was, it was clear by June 1942 that the results were disappointing and that the quality of resin oils did not meet Jupiter's standards. Consequently, the project was abandoned.

The collaboration with the Institute continued though. Brus advised Jupiter to concentrate on another technology: the

[92] Gironde Departmental Archives, Pine Institute Archives, box "Shell". N. Orlowsky, Société Anonyme des Pétroles Jupiter, Raffinerie de Pauillac: Recherches et Travaux sur la Fabrication des Huiles de Colophane, no date, p. 3.

decarboxylation of resinic acids. Instead of cracking rosin and then refining the oils, the resin acids in rosin could be decarboxylated, resulting in a range of different oils. The Pauillac engineers were in favour of the idea because the equipment used for refining cracked resin oils before could easily be adapted to treat rosins directly, reducing the number of steps in the factory. The decision proved to be successful: the new oils were lighter, and the yield better compared to cracking. By the end of 1942, a decision was taken to construct a factory capable of processing 250 tons of rosin per month, which corresponded to around 10 per cent of the entire production of the Landes Forest.[93]

It does not mean that the technology did not have its own problems. Technical challenges were numerous, including accidents in the pilot plant due to faulty materials and the inexperience of labourers who had not worked with rosin before.[94] The entire pilot plant did not satisfy any safety standards either, and air raid warnings interrupted the work on a regular basis. Despite these difficulties, Jupiter invested heavily in the new installations and hoped to establish a firm position in the lubricating oils markets. However, there were also difficulties of a formal (or even political) nature. While the question of rosin cracking was, by definition, in the scope of Jupiter's expertise, the direct use of rosin for the manufacture of lubricating oils was not. Although the scale was not identical, many local Landes companies had experimented with the technology for decades before Jupiter's engagement in the topic. The largest of them, D.R.T. (*Dérivés Résiniques et Terpéniques* (Resinic and Terpenic Derivatives)), filed a patent for its own method in 1939.[95] Brus advised the director of Jupiter's refinery,

[93] Gironde Departmental Archives, Pine Institute Archives, box "Shell". N. Orlowsky, Société Anonyme des Pétroles Jupiter, Raffinerie de Pauillac: Recherches et Travaux sur la Fabrication des Huiles de Colophane, no date, p. 7.

[94] Gironde Departmental Archives, Pine Institute Archives, box "Shell". N. Orlowsky, Société Anonyme des Pétroles Jupiter, Raffinerie de Pauillac: Recherches et Travaux sur la Fabrication des Huiles de Colophane, no date, p. 8.

[95] Gironde Departmental Archives, Pine Institute Archives, box "Shell". Letter from G. Brus to G. Laborde-Padie, 26 June 1942.

Georges Laborde-Padie, on how to avoid any potential litigation and helped him to formulate Jupiter's version of the patent that would not overlap with that of D.R.T.

This type of discussions should not be surprising between two industrial partners. What is surprising is that the director of the still existing Pine Institute Association, which held at least formally some power over the Institute, Dominique Menaut, was also one of D.R.T.'s administrators. Brus insisted in his letters to Jupiter that D.R.T. must not learn that the Institute was involved in the elaboration of a technology that was directly competing against that of D.R.T. because it would put the Institute in an ambiguous position towards the local resin industry. The Pine Institute was funded by the resin industry and was supposed to assist it, and yet it was actively collaborating with a petrochemical company that was challenging it in many other markets. Brus suggested that Jupiter should file the patent instead of the Institute.

Laborde-Padie was, however, hesitant to accept an arrangement where Jupiter would have to pay royalties to the Institute without the latter taking any legal responsibility in case D.R.T. contested the patent. Brus concurred and mentioned that this scenario was not impossible since D.R.T. was constructing its own chemical plant to manufacture lubricating oils from decarboxylated rosins, and its director was convinced he had a priority (even if Brus was sceptical of it). Ultimately, Brus rewrote the patent to make it distinct from D.R.T.'s technology, allowing it to be presented to Menaut.

Nevertheless, these negotiations were only a part of a much broader conflict that loomed on the horizon. Still, in 1942, Charles de Marcy, the U.C.R.'s director, wanted to help the local economy by creating workshops that would produce transformer oils from rosin. Brus responded that some of the oils obtained in Pauillac could serve this purpose and that the production of these oils should not be overly dispersed between small distilleries if they were to maintain consistent quality. De Marcy disagreed and explained "he would have preferred to keep the manufacture in the hands of the resin products industry". Brus related to Laborde-Padie that he "felt the old distrust of *résiniers* against *pétroliers*".[96] In his correspondence with Jupiter, the Pine Institute's director underlined that they should react immediately to prevent this "ridiculous" idea from taking root in the U.C.R. He suggested that Jupiter should invite the U.C.R. and the Pine Institute Association representatives to the refinery to show them the size of the installations and explain that similar operations would be impossible in the small local workshops in the Landes. Brus firmly believed that the collaboration with Jupiter was a technical necessity. He managed to introduce this narrative into the local press, where he explained that the partnership between the Pine Institute and the French petrochemical giant would only be beneficial to the resin industry [Vie, 1943]. It is clear that, for Brus, shared projects with Jupiter were an opportunity to test his industrial ideas on a much larger scale. He wanted to boost his authority in the French community of chemists and show that resin could be a raw feedstock as useful as petroleum.

It remains, however, less clear what the sources of the hostility towards D.R.T. were. D.R.T. was probably the most important local company and certainly showed a capacity to manage and invest in new innovative projects. A part of the answer may lie in personal conflicts. Brus developed a close relationship with Jupiter's personnel, particularly with the director Laborde-Padie and principal engineer Orlowsky, due to frequent exchanges.[97] In contrast, his relationship with D.R.T. was less cordial. In 1947, Pierre Legendre, the Institute's technical director, and another Institute researcher, together with the delegation of the local industry, travelled to the US to discover the American turpentine industry. The delegation included representatives of the U.C.R. and the D.R.T., including Jacques Boulin, the director of the company, and his engineer René

[96] Gironde Departmental Archives, Pine Institute Archives, box "Shell". Letter from G. Brus to Courtan, 24 February 1943.

[97] Gironde Departmental Archives, Pine Institute Archives, box "Shell", file "Raffinerie de pétrole Jupiter, Pauillac". Letter from G. Brus to G. Laborde-Padie, 7 October 1941; Gironde Departmental Archives, Pine Institute Archives, box "Shell". Correspondence of Orlowsky.

Escourrou. The group very quickly split into two separate teams due to personal conflicts between Legendre and Boulin. Legendre, in his letters to Brus, called Boulin "insupportable" because of his constant attacks against the Pine Institute: "[O]ne could say that he and Escourrou were entrusted with a mission to spy on us, and that they did everything to hide their own activities."[98] The situation became "ridiculous", according to Legendre, after one visit when he obtained a picture of an installation he visited: "Boulin asked me to see it, and then he kept it..."[99] Legendre also disliked the engineer Escourrou, whom he called "Mr. I know it all", and criticised him for "inappropriate behaviour" towards American women.

Brus reassured Legendre that "Boulin's character is known to everyone" and that even Dominique Menault considered him "half-insane".[100] But Legendre insisted that the problem was perhaps larger and that the U.C.R.'s leaders preferred D.R.T.'s expertise to the Institute's. He said: "I am under the impression that the U.C.R. wants to be the dictator of the Landes Forest and that it is afraid of the Pine Institute and of the role it played until this day." This is also because the Pine Institute, unlike the U.C.R., was well-known in the US, and Legendre was regularly consulted by American chemists about the future of *Bulletin de l'Institut du Pin*. Brus, in fact, explicitly encouraged Legendre to promote the Pine Institute, even at the expense of the U.C.R.

The personal conflict played a role then, but a bigger picture of local power struggles has to be taken into account at the same time. D.R.T.'s loyalty toward the U.C.R. was unquestionable, as the company was one of the organisation's pillars. The Institute, on the other hand, was independent and had considerable authority in the world of the resin industry. In other words, it could conduct its

[98] Gironde Departmental Archives, Pine Institute Archives, box "Voyage Legendre". Letter from P. Legendre to G. Brus, 18 April 1947.
[99] Gironde Departmental Archives, Pine Institute Archives, box "Voyage Legendre". Letter from P. Legendre to G. Brus, 13 April 1947.
[100] Gironde Departmental Archives, Pine Institute Archives, box "Voyage Legendre". Letter from G. Brus to P. Legendre, 22 April 1947.

own development policies and influence the local producers without passing through the U.C.R. Was the mistrust between D.R.T. and the Institute the reason the latter decided to collaborate with Jupiter? Or was it perhaps the Institute's opening towards the petroleum industry that alienated local companies? One way or another, the result was a rather complex situation in which the Institute saw external companies (be it Jupiter or Saint-Gobain) as potential saviours of the resin producing industry instead of the regional firms.

The Pine Institute carried on its research into rosin oils until at least 1946. Its work not only involved studying rosin decarboxylation and the saturation of rosin oils, but also the examination of "a series of oils manufactured by the competition".[101] By the competition, Brus meant all the companies vying with Jupiter. Between 1942 and 1946, the Institute was almost an internal laboratory of Jupiter's refinery in Pauillac.

In 1945, the collaboration was about to go one step further. Jupiter's directors agreed that a huge resin treatment installation could potentially be constructed near Pauillac, so that Jupiter could produce its own resin derivatives instead of buying rosin in the Landes. "This solution is certainly the most logical and the most economical one," explained one of Jupiter's directors. "It is, however, possible that such a solution will anger all the little producers that would see it as an attempt to replace them. This kind of reaction is so typically human!"[102] He recommended that, for the time being, Jupiter focus on advising the local companies on how to enhance the quality of rosins to demonstrate its good will. If the political situation changed and the project raised less opposition, the construction could be envisaged again. One of the directors wrote to Brus: "I imagine, dear Brus, that if we succeeded, the Landes Forest could in 20 or 30 years achieve great prosperity." He criticised, nonetheless, "antiquated customs" of the local resin producers and their

[101] Gironde Departmental Archives, Pine Institute Archives, box "Shell". Document "Travail effectué à l'Institut du Pin depuis le 31 mars 1944".

[102] Gironde Departmental Archives, Pine Institute Archives, box "Shell". Letter to G. Brus, 19 April 1945, unknown author.

faith in the power of the government to reconstruct the local economy instead of relying on their own initiative.[103] He asked Brus to contribute to the promotion of the project in the region to weaken the opposition of small entrepreneurs and workshops.

And yet, after World War II, the situation changed drastically. Only three years later, in 1948, one of Jupiter's employees wrote to Brus that Jupiter, which became Shell Française in the meantime, had no more interest in manufacturing resin oils.[104] With access to cheap petroleum on the global markets, the company had other priorities. Brus maintained some contact with the company, and in 1951, he was invited to a conference titled "Balanced economy in the forest regions" organised by Shell. Overall, however, the Pine Institute's dream to create a genuine resin-based alternative to petroleum with the help of the petrochemical industry itself was over. It is worth considering whether this outcome was inevitable. Perhaps if Jupiter had invested more money into a resin treatment facility before the end of the war, this line of studies might have continued with more success. In the US, it was the huge chemical companies, notably Hercules Powder Company, which spearheaded the development of the entire resin chemistry sector. The French failed to replicate this scenario. This led the Institute to renew its links with local stakeholders, including D.R.T., in the late 1940s.

3. Saint-Symphorien and Rosin Discolouration (1948–1952)

The post-war period brought about new challenges for resin producers. Resin came to be viewed as a low-quality ersatz from the war-time period and not a genuine alternative to petroleum-derivatives. The Pine Institute had a new mission: to demonstrate

[103] Gironde Departmental Archives, Pine Institute Archives, box "Shell". Letter to G. Brus, 19 April 1945, unknown author.

[104] Gironde Departmental Archives, Pine Institute Archives, file "Correspondance 1948, J". Letter from Shell Française to G. Brus, 7 February 1948.

that resin products could still satisfy a broad range of consumers with high-quality specialised products. One of the ways to achieve it was to make rosins brighter.

As discussed in Chapter 1, the price of rosins depended on their colour — the lighter, the better. The traditional method of brightening rosins was to expose them to sunlight in large shallow, open batches, but this discolouration method was not durable because of the resulting oxidation. The colouration was due to many factors: the quality of resin (presence of debris and water, but also meteorological and climatic factors intervening during the harvest) and the distillation methods (temperature, discontinuous or continuous, etc.), and while some improvements were made in the interwar period, there was no single solution to the problem.

After the war, in 1948, the U.C.R. made the topic a priority to legitimise its existence when the organisation came under attack. The U.C.R.'s legality had been questioned during World War II because many of its stakeholders felt that the organisation had abused its power in regulating prices. In 1943, a legal action was filed to dissolve the U.C.R., and this happened in 1948. The same year, N.U.C.R. (New Resin Corporative Union) was established to replace the old institution. The basis for the new organisation was the French labour code instead of a civil law contract, making it better grounded in the legal landscape and less prone to attacks by constitutive parties. However, the underlying problems remained. The forest owners were particularly unhappy about the distribution of profits from the resin industry in general, accusing the N.U.C.R. of favouring forest workers. Some threatened to leave the organisation, while others called it a "legal monstrosity" [Frie, 1956, p. 98]. One of the strategies to legitimise its existence among its members was to appeal to science and fund research. It was a way of showing that N.U.C.R. would be crucial for the industry's future. It is in this context that the problem of discolouration entered the organisation's agenda.

The reason for this resurgence was the knowledge brought to the region after the visit of the French delegation to the US in 1947.[105] The Americans produced half of their resinous products from dead pine trunks and not from the resin obtained by harvesting. But the rosins obtained this way were almost black and had very little market value. As a consequence, the Americans developed a range of methods to clean and brighten them. The first American patent on rosin cleaning had already been taken out in 1925.[106] It is key to point out that this American superiority came as a shock to the French. For many years, the French rosins were known for being much brighter than the American ones, due to the fact that the American industry did not clean resin before distillation. With the discolouration technology widespread in the US, this advantage was no more. The Pine Institute agreed with the N.U.C.R. that this line of research should be a priority.

In practice, two major discolouration methods were employed. The first consisted of the absorption of colouring agents with selective solvents, such as furfural (C_4H_3OCHO). This technique was popular in the US, but it required expensive installations and consumed significant amounts of solvent, making it unfeasible for smaller distilleries. The Pine Institute researchers proposed then an alternative process based on rare earths: "[T]he rosin is put in the solution with a light mineral spirit; this solution is filtrated in a plate column containing absorbing earths; the solvent is then retrieved and the earth regenerated". These earths saturated rapidly though, so the profitability of the entire endeavour depended on the costs of their regeneration. Laboratory tests were very promising, and the N.U.C.R. decided to fund the project on a semi-industrial level. One local company made its distillation workshops in a little village, Saint-Symphorien, available for tests. The Pine Institute and another firm

[105] Gironde Departmental Archives, Pine Institute Archives, file "Essais Usines". P. Legendre, G. Brus, "Rapport sur la création de l'usine pilote de décoloration des brais et colophanes de St Symphorien", 17 November 1949.

[106] Gironde Departmental Archives, Pine Institute Archives, file "Décoloration des colophanes". A. R. Autrey, Procédé de décoloration et raffinage des gemmes et des produits résineux, patent publication date: 27 October 1925, patent number: U.S.P. 1.559.399.

were responsible for the technical control. The key questions to answer concerned the optimal dimensions of columns, as well as the minimum improvement of the yield and of the colour grade necessary to make the project profitable.

In 1949, several issues arose regarding the discolouration project. The installation was under-financed and too small, and the equipment for rare-earth regeneration was not efficient, resulting in an inability to evaluate overall efficiency. The Pine Institute explained in its first activity report to the N.U.C.R. that "this pilot plant should be considered a testing device, the development of which will require time and modifications".[107] Both Brus and Legendre tried to diplomatically convey that it would be an error if the local entrepreneurs thought that the technology was in their immediate reach and that it was on the point of transforming the entire industry. On the contrary, they explained, it could take many years before it would be ready. In April 1950, the tests remained inconclusive. Technical problems caused "pessimism" among the chemists working in the plant.[108]

In September 1950, the N.U.C.R. reassured its members that the first results were satisfying and that new installations were being constructed [Bonnin, 1950]. The report from June 1951 mentioned some ongoing difficulties, but Legendre argued that the project should not be abandoned because the discolouration itself was very efficient, and the technical problems were not insurmountable.[109] In July 1951, one external consultant suggested a few improvements in the composition of substances used for filtration but indicated that he believed many more tests were necessary to optimise the process.

---

[107] Gironde Departmental Archives, Pine Institute Archives, file "Essais Usines". "Rapport sur les essais faits à l'usine pilote de décoloration des colophanes de St-Symphorien, essais des 29 et 30 November 1949".

[108] Gironde Departmental Archives, Pine Institute Archives, file "Essais Usines". "Rapport sur l'usine pilote de décoloration des brais et colophanes de St-Symphorien, Visite effectuée avec M. Courret le 28 April 1950".

[109] Gironde Departmental Archives, Pine Institute Archives, file "Essais Usines". "Visite effectuée à l'usine — pilote de Saint-Symphorien par MM. Legendre et Grancher sur demande de M. Poudenx", report from 25–26 June 1951.

In April 1952, a new internal report was optimistic and claimed that the yield was good and that the only issue was that iron and copper in the piping coloured rosins, so they had to be entirely replaced with stainless steel.[110] And yet, this 1952 report is the last one concerning the Saint-Symphorien plant in the Institute's archives. The N.U.C.R remained silent on the topic as well. It appears that after no less than three years of investment, the financial return of the project was null, and the project was abandoned, although it is possible that some technical knowledge was transmitted to other companies and some know-how on resin discolouration was kept in the region.

4. SOLARED, Fréjacques Commission, and Rosin Hydrogenation

At the beginning of the 1950s, when the Saint-Symphorien project still held some promises, Brus and local entrepreneurs wanted to ensure the future of industrial research in Aquitaine. They envisaged the construction of a huge semi-industrial workshop in the image of the American laboratory in Olustee in Florida.[111] It was meant to not only have strictly technical personnel but also a "techno-commercial" staff devoted exclusively to assisting manufacturers in the region. Early plans suggested that the N.U.C.R. and Pine Institute would jointly supervise this facility, but the project was eventually abandoned. Nevertheless, this effort demonstrated a recognition that the situation of the French resin industry on global markets was declining.

In the 1910s to 1920s, the French produced around 20,000 tons of turpentine spirit annually, which roughly corresponded to 20 per cent of global production. Up to 70 per cent was manufactured in the US, and the remaining 10 per cent in other countries. However,

[110] Gironde Departmental Archives, Pine Institute Archives, file "Essais Usines". "Rapport sur les essais", 10 April 1952.

[111] Gironde Departmental Archives, Pine Institute Archives, box "Divers". P. Legendre, "Sur les dérivés des produits résineux", no date.

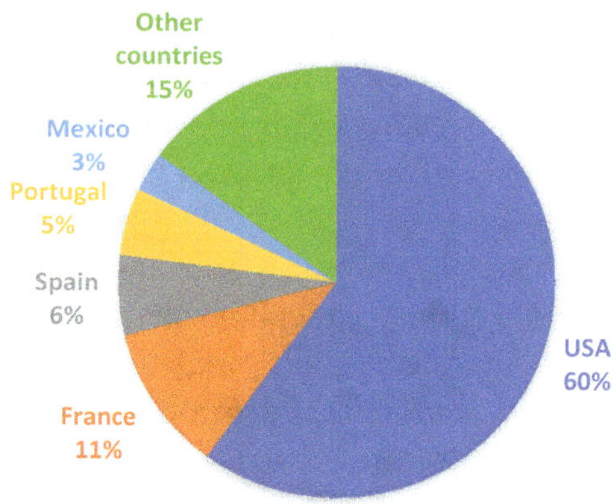

Figure 29: Global turpentine spirit production in 1952.[112]

by the 1950s, a major shift was under way as more and more developing countries joined the competition, threatening the leading stakeholders (Figure 29).

In 1952, French production barely exceeded 16,000 tons, 20 per cent less than before the war, while new competitors, such as Spain and Portugal, but also Italy and Greece on a smaller scale, as well as East-Asia, were growing in importance. The future looked bleak for France. In the US, much cheaper wood turpentine spirit extracted from pine stumps was gaining an increasingly prominent position, and in other countries, the labour costs were considerably lower, pushing resin prices down. At the same time, turpentine substitutes were becoming more and more prominent. Not only were spirits from petroleum distilled with higher standards, but a range of pine spirits coming from Scandinavian paper mills were invading the French market.[113] All of this was disquieting for the inhabitants of the forest regions. There was, in fact, a widespread conviction that: "turpentining was practically indispensable for the maintenance of

[112] Established from data in: Gironde Departmental Archives, Pine Institute Archives, file "Solared". G. Brus, "L'industrie chimique et les produits résineux", Congrès National du Bois 1953–1954.

[113] Gironde Departmental Archives, Pine Institute Archives, file "Solared". G. Brus, "L'industrie chimique et les produits résineux", Congrès National du Bois 1953–1954.

the Landes Forest that provides, in the form of wood, an important contribution to the French economy."[114] It was believed that the presence of forest workers was necessary to prevent fires and that without them the entire forest would probably perish, leading to disastrous economic consequences for the entire nation, even though in the 1950s an increasing number of forest owners questioned this position [Frie, 1956, 27].

In the middle of the 1950s, the N.U.C.R.'s lobbying efforts paid off, and the government took more interest in the resin industry and its development. The Commission for Studies and Development of Resin Products, called, in short, the Fréjacques Commission due to the name of its president, was established with the support of the Ministry of National Economy, the Ministry of Trade and Industry, and the *Conseil supérieur de la recherche scientifique et du progrès Technique* (High Council of Scientific Research and Technical Progress) — an influential national science policy body.[115] This commission initially had five members: Maurice Fréjacques, scientific director of Pechiney (French chemical and electrometallurgical giant), Dupont, professor at École normale supérieure, Brus, the director of the Pine Institute, André Borocco, general director of the chemical plants of Saint-Gobain, and M. Castets, general director of Organico (future Aquitaine-Total-Organico). A certain Echars from the government was soon added to this list, and especially Henri Longchambon, the president of the High Council of Scientific Research and Technical Progress.[116] It is clear that the commission reunited some of the leading French scholars and the representatives of the largest French chemical companies.

One of the first conclusions of the Fréjacques Commission was that the research on resin applications was underfunded in the region and that the industry should spend more on innovation.

[114] Gironde Departmental Archives, Pine Institute Archives, file "Solared". Note on "Commission d'étude du Développement des Produits Résineux".
[115] Note on "Commission d'étude du Développement des Produits Résineux", Gironde Departmental Archives, Pine Institute Archives, file "Solared".
[116] Gironde Departmental Archives, Pine Institute Archives, file "Fréjacques I". Letter from M. Fréjacques to G. Brus, 6 May 1955.

Another conclusion was that the quality of resinous products remained very inconsistent and that addressing this challenge should be a priority. To achieve technical improvements on a larger scale, it was suggested to group dispersed workshops.[117] In fact, in 1955, there were 77 independent distillation workshops in the Landes Forest. The commission wanted to reduce this number to at least 25. Finally, the commission underlined the necessity to develop new markets, notably for polymerised, disproportionated, and hydrogenated rosins, as well as for chlorinated resin-based insecticides, which apparently held many promises. It remarked that while some of the products mentioned above have been patented in different countries, the vast majority of them were initially developed in the Pine Institute's premises, and the French industry should make use of them without fear of legal difficulties. The commission entrusted the N.U.C.R. with the development of a broader development plan, and Legendre, the Institute's technical director, was to prepare a report on the possibility of constructing middle-size process plants in the region.

Following months of deliberation and generating numerous reports on the state of scientific and industrial progress in the Landes resin industry, the Fréjacques Commission decided on a clear direction. It called for the construction of a semi-industrial plant to study the manufacture of polymerised rosins and maleopimaric acid. According to the reports, there was a market for at least 2,000 tons of polymerised rosin. As for maleopimaric acid, there was no market yet, but if it was obtained directly from the gum, it would become cheaper and could potentially find uses in the varnish industry that occasionally manufactured the maleopimaric acid on its own from rosin.[118] In other words, the Institute tried to put in

---

[117] Gironde Departmental Archives, Pine Institute Archives, file "Fréjacques I". Commission d'étude du développement de l'emploi des produits résineux, "Compte rendu de la 1ère réunion".

[118] Gironde Departmental Archives, Pine Institute Archives, file "Solared". "Note sur la commission d'étude du Développement des Produits Résineux".

action some of its major innovations: polymerised rosins and direct gum fractionation. A new trade company was established in 1955 to build the semi-industrial installation: *Société Landaise des Résines et de leurs Dérivés* (the Landes Society for Resins and their Derivatives — SOLARED) in Mont-de-Marsan. Its share capital reached one million francs, and among its shareholders were the largest local companies,[119] with the exception of the most important, D.R.T., which had been given a different mission from the Fréjacques Commission: to construct a workshop manufacturing pinic acid from turpentine spirit.

SOLARED's founding statute outlined its objectives: the manufacture and research of new resin-based products, as well as intellectual property management.[120] The statute explicitly stated that the Pine Institute would carry out regular quality control check on all substances produced by the company. This provision was not a coincidence. In fact, Brus and Legendre required this fragment to be added, as they explained that the Institute could not officially endorse any industrial initiative if it "does not adopt, in all honesty, modern technical methods". And for that to happen, the Pine Institute should guide the company's activities. For Brus and Legendre, this was a matter of prestige. If the Pine Institute officially supported an initiative that was not scientifically and technologically sound, "this would be the end of the Pine Institute", they explained.[121]

---

[119] Etablissements Veuve Joseph Farbos et Huileries des Landes Réunies, Société Huileries et raffinerie de Saint Symphorien, Société Charles Navarre, Société Industrielle des Résines, Société Industrielle des Bois et Résineux, Etablissements Pierre Maydieu. See: Gironde Departmental Archives, Pine Institute Archives, file "Fréjacques II". Commission d'étude du développement de l'emploi des produits résineux, "Compte rendu de la deuxième rencontre à Bordeaux le 30 June 1955".

[120] Gironde Departmental Archives, Pine Institute Archives, file "Fréjacques II". Commission d'étude du développement de l'emploi des produits résineux, "Compte rendu de la deuxième rencontre à Bordeaux le 30 June 1955".

[121] Gironde Departmental Archives, Pine Institute Archives, file "Solared". "Résumé de la conversation entre G. Brus et P. Legendre du 6 mai 1955".

The level of mistrust towards the level of expertise of the local companies may seem surprising, but it reflects the climate reigning at the time. The Institute's supervision was confirmed in a convention between SOLARED and the N.U.C.R., which explained that the "semi-industrial installation [will be] managed and administrated by SOLARED, but placed under the scientific direction of the Pine Institute".[122] More interestingly, the convention explained that all the equipment bought from the N.U.C.R.'s subventions to SOLARED would remain the N.U.C.R.'s property. However, all the patents developed by SOLARED (no matter the source of funding) would be filed by the Pine Institute. Not only would the Institute control SOLARED's manufacturing operations, it would also acquire all intellectual property generated through its activities. In other words, SOLARED would become an extension of the Institute, a semi-industrial research centre funded by the industry. The Institute would have even retained an influence on strictly trade-related decisions concerning the commercialisation of the invented products. In short, had the SOLARED plant been constructed, it would have started a new golden age of pine chemistry research in Aquitaine.

However, the project quickly encountered financial hurdles. The initial cost of the plant was valued at 33 million francs. The N.U.C.R. itself decided to invest 5 million francs immediately and committed to collecting the remainder from the local companies, the Ministry of National Economy, and the High Council of Scientific Research and Technical Progress.[123] On 30 November 1956, a meeting of the Fréjacques Commission took place, in which it was announced that SOLARED still had not managed to secure basic funding, even after one year after the initial talks. The Ministry of Economy explained that the government could cover up to 50 per cent of total costs, but the N.U.C.R. and the local industry needed to find the other half

[122] Gironde Departmental Archives, Pine Institute Archives, file "Solared". "Projet d'une convention entre la SOLARED et l'U.C.R".

[123] Gironde Departmental Archives, Pine Institute Archives, file "Solared". "Projet d'une convention entre la SOLARED et l'U.C.R".

first.[124] Not only that, some corrections to the company's strategy were also suggested. For instance, a D.R.T. representative indicated that before investing so much in a plant designed to find new methods for the polymerisation of rosins, some polymerised rosin should be manufactured beforehand in the existing distilleries and sent around to conduct a more thorough market study on whether it could really sell.[125] This would help to assess the profitability of the entire endeavour, as there had been no preliminary market studies conducted. It was suggested that these could perhaps be conducted with similar foreign rosins, but the objection was that this would not be possible because polymerised rosins produced in the Landes were not identical to the ones from abroad. This shows again a broader issue concerning the importance of regional differences in the manufacture of resinous products.

What was the fate of SOLARED? In an activity report for 1957–58, the Pine Institute's director noted that dismuted and polymerised rosins were being manufactured by two companies with the Institute's assistance.[126] In the following year, he mentioned another one, but this was the last time the topic was ever discussed in the reports.[127] We can surmise that if SOLARED had been a success, it would have been featured more prominently in the Institute's reports or at least in the N.U.C.R.'s annual reports. Everything indicates that SOLARED failed to get funding, but the technological trajectories it was supposed to develop were not abandoned, and the Pine Institute tried to save them through new collaborations. Interestingly, the failure of SOLARED coincided with the dissolution of the

[124] Gironde Departmental Archives, Pine Institute Archives, file "Solared". "Commission n°29-2, Section Développement des Produits Résineux, Procès-Verbal de la Réunion Tenue à Paris le 30 November 1956".
[125] Gironde Departmental Archives, Pine Institute Archives, file "Solared". "Commission n°29-2, Section Développement des Produits Résineux, Procès-Verbal de la Réunion Tenue à Paris le 30 November 1956".
[126] Gironde Departmental Archives, Pine Institute Archives, file "U.C.R". "Résumé des travaux effectués par l'Institut du Pin au cours de la campagne 1957–1958".
[127] Gironde Departmental Archives, Pine Institute Archives, file "U.C.R". "Résumé des travaux effectués par l'Institut du Pin au cours de la campagne 1958–1959".

Fréjacques Commission in 1958 [Dardel, 2006]. It is also possible that, the Ministry of National Economy refused to support the project, leaving SOLARED with less than half of the estimated basic funding. One way or another, this last great regional scientific and industrial initiative on resin chemistry was over.

5. Resins and Space Conquest: Véronique Rockets

The interest in resin derivatives went, however, far beyond the regional industry — probably the most ambitious project involving the Pine Institute concerned the use of turpentine in rocket fuels in the French national space programme. This endeavour dates back to May 1946, when the *Laboratoire de recherches balistiques et aérodynamiques* (Laboratory of Ballistics and Aerodynamics Research — L.R.B.A.) was established. A total of 123 German engineers worked there with French scientists on the construction of the first French rocket, Véronique [Huwart, 2000]. Rocket fuels need to get oxygen from another substance called a propellant. In the 1950s, the number of oxidising agents was relatively limited — hydrogen peroxide, fluorine, tetranitromethane, and especially liquid oxygen and nitric acid [Warren, 1958, p. 12]. The choice of fuels was much larger. These were typically divided into four categories: (1) ammonia, hydrazine, metallic hydrides; (2) amines; (3) alcohols; and (4) petroleum derivatives (spirit, kerosene). In 1954, French engineers, after having tested various combinations of fuels and propellants, finally settled on the combination of kerosene and nitric acid. However, the results were not satisfactory, and the L.R.B.A. came under fire to justify escalating costs [Varnoteaux, 2003].

What brought the project back into the spotlight was the launch of Sputnik, prompting the Western Bloc to put more effort into rocket research. In the US, the National Aeronautics and Space Administration was created in 1958. In France, the Space Research Committee came into existence in 1959, alongside the *Société pour l'étude et la réalisation d'engins balistiques* (Society of Study and Realisation of Ballistic Devices — S.E.R.E.B.), whose primary mission was to develop means of transporting atom bombs [Varnoteaux,

2003]. These new initiatives led, on the one hand, to the development of the Mirage aircraft, and on the other hand, to the renewal of interest in Véronique. In a bold move, the L.R.B.A. decided to give up on kerosene and turned to a cheap national fuel: pine turpentine spirit. The first tests showed promising results and French rocket scientists affirmed that the use of turpentine spirit eliminated many of the difficulties encountered with kerosene [Varnoteaux, 2002, p. 288]. These first attempts, which took place in the deserts of the Sahara on the Algerian territory, immediately excited the resin producers of the Landes [N. N., 1959].

It is unclear when exactly the L.R.B.A. entered into contact with the Pine Institute. Brus was certainly closely tied to military circles for a long time. In 1937, he was responsible for the establishment of chemical gas detection facilities in Bordeaux in case of air raids, and he was an influential figure when, for a brief moment, the city was the seat of the French government during the war.[128] He also worked on the varnishes for military aircraft and had a turbulent collaboration with the French Ministry of Defence on camphor. In 1945, Brus intervened at a French Aviation Congress on the possibility of manufacturing high-octane fuels from resinous products.[129]

Whatever the exact circumstances of the initial contact were, the collaboration with L.R.B.A. was probably the most prestigious in the history of the Pine Institute. In 1961, at the request of L.R.B.A., the Institute led "in-depth studies on the composition of turpentine spirit using vapour phase chromatography". This research concentrated on "the identification of minor components" that might affect the quality of the fuels.[130] As usual, the Institute advertised its studies in *Bois et Résineux*, promising "new important market outlets"

[128] Gironde Departmental Archives, Pine Institute Archives, file "Correspondance 1937 G. B., I". Letter from the director of Bureau d'Hygiène to G. Brus, 6 June 1937.

[129] Gironde Departmental Archives, Pine Institute Archives, file "Correspondance 1945, C". Letter from the organizers of the Congrès national annuel de l'Aviation française to G. Brus, 15 March 1945.

[130] Gironde Departmental Archives, Pine Institute Archives, file "U.C.R.". "Travaux de l'Institut du Pin, Campagne 1961–1962".

[Rusterholz, 1962]. This line of studies was conducted with the help of an L.R.B.A. engineer detached to the Institute in 1962 [Rusterholz, 1964]. That same year, an L.R.B.A. delegation came to Bordeaux to discover the distillation methods practised in the region, in particular, to explore the possibility of manufacturing heavier spirits.

The result was an internal report on the chemistry of resins and on the technical challenges surrounding their treatment. It notably included an exact description of the Pine Institute's continuous distillation column. This suggests, however, that the L.R.B.A. did not want to buy resinous products from the local producers but expected to manufacture them on their own to ensure their quality and composition. In 1963, the Institute studied some samples for the L.R.B.A. and was kept in the loop about the challenges encountered in the preparation of the turpentine-based rocket fuel. The Institute's archives kept extensive documentation on the launch tests.[131] It seems that after 1963, the brief collaboration was over, and the Pine Institute was not involved in rocket fuel studies anymore. In total, three models of Véronique (AGI, 61 and 61M) were fuelled with turpentine spirit, and 71 rockets using it were launched between 1959 and 1971. In addition, in 1962, the L.R.B.A. was entrusted with the construction of ten rockets more powerful than Véronique: Super-Véroniques and Vestas. Four launches took place, all of them using turpentine spirit. In 1965, one of the modules of the Diamant rocket that placed the first French satellite in orbit also used the Landes fuel [Huwart, 2002, p. 385].

How can the brief collaboration of the Institute on rocket fuels be evaluated retrospectively? Three perspectives are possible. First, the role of the Bordeaux institution should not be overestimated. The first turpentine-fuelled Véronique had been launched in 1959, prior to the collaboration with the Institute. The military engineers turned to the Institute to improve the fuel, not to invent it. Furthermore, the collaboration itself took place between 1961 and 1963, only a fraction of the total history of the project. Not only that, many studies on turpentine were also conducted by the laboratory of organic chemistry of the famous French military school *École polytechnique*

[131] Gironde Departmental Archives, Pine Institute Archives, file "L.R.B.A.".

in Paris.[132] Without diminishing the contributions of the Pine Institute, its role in the French space programme was limited.

The second perspective takes into account the place of the Institute in the research infrastructure of the Bordeaux region. In fact, the S.E.R.E.B. developed much of its activities precisely in Aquitaine due to the presence of an ancient powder factory in Saint-Médard-en-Jalles [Gossot and Peuron, 2007]. Already in the 1960s, a range of new aerospace and rocket science laboratories was established in the Landes.[133] All of them closely worked with different laboratories of the University of Bordeaux. While the S.E.R.E.B. and Landes aerospace laboratories did not work directly with the Pine Institute, the latter was, through its collaboration with the L.R.B.A., part of the diverse range of stakeholders that developed expertise in aerospace research, which has contributed to the Aquitaine region's reputation in this field [Maison-Soulard and Frigant, 2020].

Finally, from the resin industry's point of view, the Véronique project was of tremendous symbolic importance. Resin became seen as a serious competitor of petroleum derivatives in the most prestigious application. This gave hope that turpentine could yet reassert its place in the petrochemical era.

Although the history of turpentine as a rocket fuel ended in France with the Véronique project, it is not over yet at a global level, as turpentine has been used in the US by the private space industry [Antunes, 2012, p. 57].

6. Institutional Architecture of the Laboratories of the Pine Institute and their Scientific Production after the War (1948–1969)

How was the Institute organised after World War II? Reports are scarce, but the ones that exist reveal that in 1949, the Institute was

[132] Gironde Departmental Archives, Pine Institute Archives, file "L.R.B.A.". "Travaux exécutés par le Laboratoire du Professeur Baranger".

[133] *Centre d'essais au sol des propulseurs* (1961) that became *Centre d'achèvement et d'essais des propulseurs et engins* in 1967; *Société d'étude de la propulsion par réaction* (1963); *Centre d'essais des Landes* (1965).

structured around nine sections: (1) Direction and administrative services; (2) Laboratory of research on chemistry and the use of resinous products and their derivatives; (3) Laboratory of research on wood and its derivatives; (4) Laboratory of physical research; (5) Forest biology laboratory; (6) Analytical service; (7) Documentation service; (8) Teaching service; and (9) Technical consultancy commissions (Table 8).

A few changes in comparison to the interwar period are evident; for example, the turpentine, rosin, and gum harvesting laboratories were merged into a single more general resin research laboratory. There were also newcomers — small laboratories of physics and biology. More interestingly, the last unit, technical commissions, was created to answer the immediate needs of the companies requiring expertise on a specific type of issues. There were a few of them: on gum, on resin derivatives, on paints and varnishes, on synthetic resins, on soaps and detergents, on cables, etc. Their goal was to maintain a permanent link between scientists and industrialists in the region.

Table 8: The evolution of the organisation of the Pine Institute over the years. n — new services created in the 1940s

Initial vision in 1921	Organisation between the 1920s and 1940s	Organisation between the 1940s and 1960s
1. Research laboratory (four divisions: gums, turpentine, rosins, and wood)	1. Harvesting and treatment of gum resin lab	1. Resinous products and their derivatives lab (result of a merger of resin, turpentine and rosin labs)
	2. Turpentine spirit, terpenes and derivatives lab	
	3. Rosins or derivatives lab	
2. Technical teaching	4. Wood and paper lab (with two subsections: (i) wood distillation/tar, and (ii) paper)	2. Wood and its derivatives lab
3. Analytical service		3. Physics lab (n)
4. Documentation office		4. Forest biology lab (n)
5. Industrial office	5. Analytical service	5. Analytical service
	—	6. Documentation service
	Documentation, teaching, and consultancy were provided but not formalised	7. Teaching service
		8. Technical commissions (n)

The exact number of employees of the four main laboratories (resin chemistry, wood chemistry, forest biology, physics) remains unknown, but they were not equal in size. The forest biology laboratory was the smallest. It was established in 1948 and had most likely only one permanent member: Professor Roger David. Professor David was a member of the prestigious French *Académie des Sciences* (Academy of Sciences) and one of the greatest specialists in pine biology. He wanted to reinvigorate the pine biology research in the Institute in close collaboration with local foresters, and his studies concerned the impact of forest clearings on the productivity of pines as well as the development of hormones for the elimination of the undergrowth to reduce fire risks. The study of the latter topic became a priority following a tragic forest fire in 1949 that ravaged 100,000 hectares of the Landes Forest and took 82 lives [Deville, 2009]. The Pine Institute presented itself as a key stakeholder in the battles against forest fires and obtained a grant of 300,000 francs for studies on fire prevention using chemical means. David was also a pioneer in the field of pine seed germination and worked on selective breeding of higher and faster-growing pines [N. N., 1949b]. In the 1950s, he was responsible for the introduction of gum harvesting with acid in the region.[134] In total, between 1948 and 1959 (the years of the laboratory's activity), only eleven articles on pine biology were published. It seems that its contributions were above all, practical, and, after the closure of major projects, the laboratory discontinued its activities. It is worthwhile to mention that the closure of the forest biology laboratory in the late 1950s marked a definitive disengagement of the Institute from forest-related studies — a line of research that was of the Institute's marking features in the interwar period.

In contrast, the Institute's wood and derivatives laboratory had a different dynamic. In the interwar period, the Institute had a wood and papermaking laboratory, but it held a less prominent role in the Institute's structures. Back then, wood chemistry specialists worked

[134] Gironde Departmental Archives, Pine Institute Archives, box "Divers". G. Brus, "Rapport sur le fonctionnement et les travaux de l'Institut du Pin, année 1956".

on forest fuels on the one hand and on the use of pine wood in the paper mills on the other. This second topic took precedence and became more and more important throughout the 1950s and the 1960s, thanks to two scientists, Pierre and Huguette Monzies, a married couple employed by the Institute. Their work was devoted to the study of the properties of pine-based paper pulp. In 1958, the Institute's wood laboratory became one of the founding units of the *Centre technique des papiers, cartons et cellulose* (Technical Centre of Paper, Cardboard, and Cellulose), a nation-wide entity coordinating paper-related research on a national scale.[135] It would later become the *Centre Technique du Papier* (Technical Centre for Paper), which remains a key institution in the field until this day.

Monzies' laboratory was highly regarded in the region and received substantial funding from local companies. In 1966, the local industry awarded the laboratory with a large subvention of 528,150 francs to cover the costs of high-quality equipment that was meant to serve the entire sector. Between 1947 and 1967, about twenty publications and oral communications concerning papermaking were issued by the laboratory, with more than half of them in the 1960s, which shows the dynamic development of the branch.

Overall, however, both forest biology and paper remained marginal topics for the Institute, with few papers having been published overall. It was the Laboratory of Research on Chemistry and Uses of Resinous Products of the Pine Institute (Table 8) that employed the vast majority of the Institute's scientists. The laboratory was directed from 1949 by a Vietnamese chemist, Lê Van Thoi.

Lê Van Thoi was born in 1917 in a poor agricultural district of Gò Dầu in the south of Vietnam (Figure 30).[136] Thanks to a French government scholarship, he got the opportunity to study in France,

[135] Gironde Departmental Archives, Pine Institute Archives. R. Bentejac, "Historique de l'Institut du Pin et de ses activités dans le domaine de chimie", 1996, p. 7.

[136] Office of the Foreign Secretary. National Academy of Sciences, Annual Report of the Board of Science and Technology for International Development, 1973, p. 12; Article about Lê Van Thoi on the website of the newspaper of the Vietnamese communist party (https://fr.nhandan.vn/le-van-thoi-le-scientifique-qui-sest-distingue-pour-son-savoir-hors-pair-post72375.html, last accessed 3 April 2023).

Figure 30: Lê Van Thoi (1917–1983). Vietnamese chemist and the director of the Pine Institute's main laboratory between 1949 and 1956.
Source: Archives of *Le Courrier du Vietnam*[137]

where he graduated with degrees in chemistry and physics in 1942. In 1947, he obtained his Ph.D. under the direction of Brus at the Pine Institute. Two years later, in 1949, the young researcher became the section head of a new large Laboratory of Chemistry and Utilisation of Resinous Products and Their Derivatives that replaced the two laboratories of turpentine and rosin, which had existed during the interwar period. Lê Van Thoi left the Pine Institute in 1956 and then, in 1958, went back to Vietnam. On his return to Vietnam, he became the director of the Department of Sciences of the University of Saigon and the director of the Vietnamese General Department of Atomic Energy. He was also the president of the scientific council of South Vietnam after the war.

In 1949, the first year for which there is detailed data after the war, Le Van Thoi's laboratory was made up of eight chemical engineers,

[137] In the public domain. Courtesy of Nhân Dân (https://fr.nhandan.vn/le-van-thoi-le-scientifique-qui-sest-distingue-pour-son-savoir-hors-pair-post72375.html, last accessed 3 April 2023). Also available in the Pine Institute Archives, file "Photos".

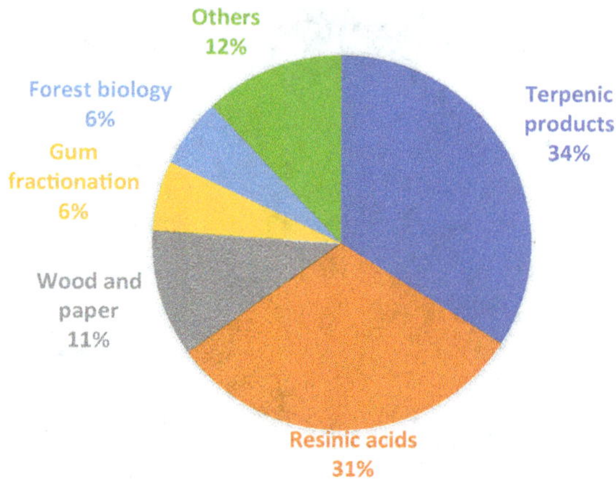

Figure 31: Major themes of the Institute's publications (1947–1969).

two assistants, three Ph.D. students, one "mechanist", one storekeeper, and three cleaning and maintenance personnel — in total, 18 employees, more than the entire Pine Institute's staff in the first years of World War II [N. N., 1949b]. This laboratory was assisted by a small laboratory of physical research of Professor Auguste Rousset, one of the most prominent proponents of Raman spectroscopy in France.[138]

Out of 190 articles and oral communications originating from the Pine Institute in 1947–1969, around 65 per cent concerned turpentine spirit and rosin. The exact proportion of different topics is displayed in Figure 31.

It can be observed that, forest biology and paper studies aside, an important cohort of articles concerned the new mode of gum fractionation actively advocated by Brus. However, the two biggest research themes were the same as in the interwar period: turpentine and rosin, as well as their derivatives. At first glance, it may seem that these two topics remained in a relative balance. However, this impression is clarified if one looks at each theme separately (Figure 32).

Figure 32 shows that turpentine spirit was completely discarded as a research topic after the war. It was certainly still a valuable

[138] Biographical note on Auguste Rousset (http://christian.j.borde.free.fr/AugusteRousset.pdf, last accessed 3 April 2023).

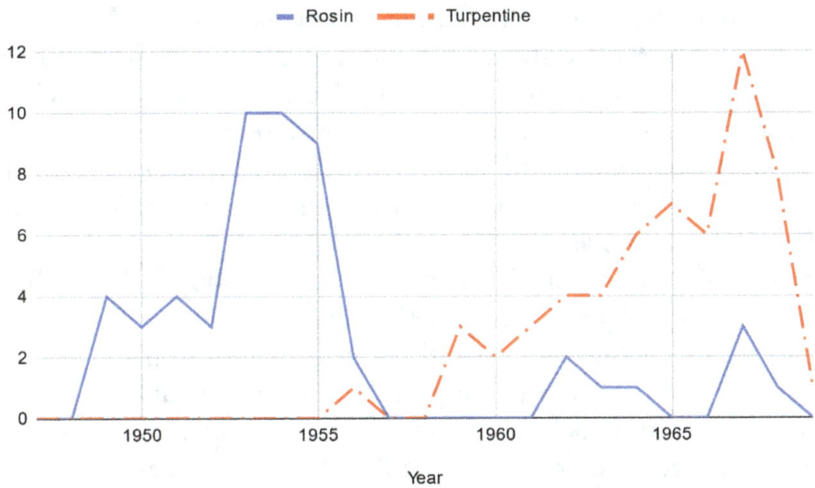

Figure 32: Number of papers on rosin and resinic acids and on turpentine spirit and its derivatives published by the Pine Institute researchers.

product, but with the novel perspectives opened by gum fractionation, as well as by the rosin hydrogenation and dehydrogenation, resin acids were expected to impact a number of new markets. As a consequence, the numerous studies were conducted in the 1940s and early 1950s to explore the properties of different individual acids. In 1953 alone, as explained in an internal report, "45 new derivatives have been prepared from dehydroabietic acid, 23 from dextropimaric acid, 40 from dihydrodextropimaric acid."[139] The synthesis of new compounds was the driving force of all the activities of the Pine Institute and yet, just a few years later, resin acids disappeared from the Institute's agenda and turpentine retook its former position. Three reasons could explain this shift.

First, the failures of the Saint-Symphorien and SOLARED projects were a harsh experience for the Institute's researchers. Neither discolouration nor hydrogenation managed to transform the industrial landscape of the region. As for polymerisation, it was practised on a

[139] Gironde Departmental Archives, Pine Institute Archives, file "U.C.R.". "Institut du Pin", no date.

small scale by bigger companies, but it remained marginal overall. It is possible that these difficulties disillusioned the Institute's research about resin acids' true potential and discouraged further research.

The second reason was the scientific progress concerning turpentine and its derivatives. Already in 1953, the Pine Institute filed a patent titled "New Polyamides" for the manufacture of a substance called Terlon, a polyamide fibre obtained by the polycondensation of diamines with pinic acid, pinene oxidation product.[140] Visually, it was similar to Nylon, and in 1954, Brus and Legendre hoped it might become a big market outlet for turpentine spirit. The pinic acid seemed to offer more and more opportunities, unlike the resin acids from rosin, which had proved difficult to commercialise. The 1958 annual research report of the Pine Institute explained that:

> [T]he synthesis and the properties of different esters of pinic acid were studied because of the importance these esters are susceptible to have, not only as plastifiers of some synthetic resins, but also due to their interest, given their stability and their viscosity curve, in the lubrication of turbo-reactors and automatic weapons, as well as control fluids for servomotors [Rusterholz, 1958].

In 1959, in parallel to polyamides, the Institute studied hydroperoxydes (pinane, menthane, carvomenthene, limonene), which could potentially serve as polymerisation catalysts, in particular for the manufacture of synthetic rubber. Another promising topic was the study of nopinene polymers and the polymerisation of vinyl derivatives.[141] Polymersation, initially started in the context of rosins, was a rapidly growing study theme in the 1960s in the context of turpentine derivatives. Finally, the collaboration with the L.R.B.A. on turpentine fuels boosted the self-esteem of the Institute's scientists and improved their perception of turpentine.[142]

---

[140] Gironde Departmental Archives, Pine Institute Archives, file "U.C.R.". "Institut du Pin", 1954.
[141] Gironde Departmental Archives, Pine Institute Archives, file "U.C.R.". "Travaux effectués par l'Institut du Pin au cours de la campagne 1962–1963".
[142] Gironde Departmental Archives, Pine Institute Archives, file "U.C.R.". "Travaux effectués par l'Institut du Pin au cours de la campagne 1959–1960".

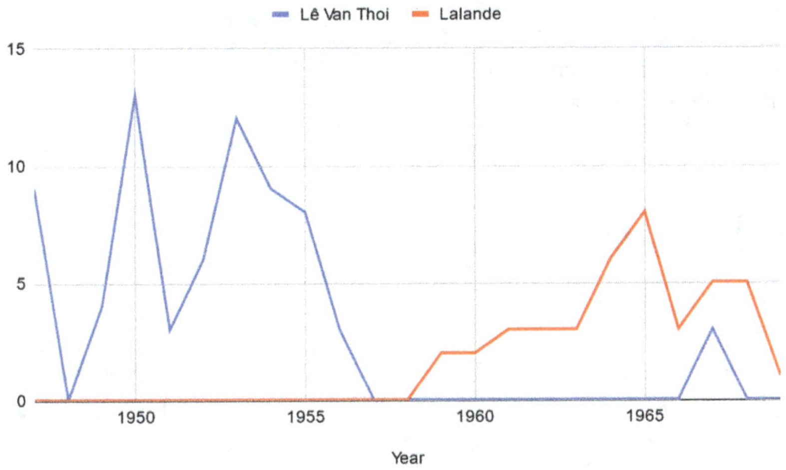

Figure 33: Number of papers of Lê Van Thoi and Lalande (1947–1967).

The third reason for the marginalisation of resin acids and the return of turpentines was of a more personal nature. The publication curve of the resin laboratory chief and the Institute's most productive scientist, Lê Van Thoi (70 co-signed articles), showed a pattern similar to that of rosins. A specialist in rosin acids, his departure in 1956 influenced the entire research programme of the Institute. The following year, in 1957, Brus replaced him with Roger Lalande, a chemist trained in terpene chemistry. Lalande's publication curve overlapped perfectly with turpentine's. On the whole, both laboratory chiefs influenced the choice of topics studied in the Pine Institute laboratories (Figure 33).

What is the overall productivity of the Pine Institute's researchers over this period? Table 9 attempts to answer this question by identifying the most productive researchers alongside their topics of choice. Lê Van Thoi and Lalande were the unquestionable top performers in terms of the number of publications (possibly because they were section heads and co-signed most of the articles). Despite having many other functions, Brus also had an impressive publication score. It is worth noting that Brus was also the dean of the Bordeaux Faculty of Sciences until 1961, a position he had to relinquish due to health problems [N. N., 1963]. Interestingly, Legendre, the Institute's technical director and its "face" towards the outside

Table 9: Most productive researchers of the Pine Institute (1947–1969)

Researcher with at least 10 publications	Number of publications and communications	Percentage of total (187)	Focus
Lê Van Thoi	70	37.4%	Resin acids
Roger Lalande	41	21.9%	Terpenes
Georges Brus	37	19.8%	Rosin, turpentine, wood and papers
Mlle François	25	13.3%	Resin acids and terpenes
Pierre Monzie	21	11.2%	Wood and paper
Pierre Legendre	16	8.6%	Resinic acids and gum
Jean Moulines	13	7.0%	Terpenes
Ourgaud	12	6.4%	Resin acids
André Belloc	10	5.3%	Resin acids

world, published only 16 articles between 1947 and 1969, much fewer than the resin chemistry giants, such as Brus and Dupont, during their respective tenures as the Institute's directors. He did, however, contribute to consulting activities throughout his career, not to mention the patents for which there is no exact data.

Conclusions: Identity Under Construction

Following World War II, the Pine Institute sought to reposition itself in the local industrial landscape. The situation of the industry was deteriorating and the new policy of the Institute was to focus on large-scale industrial collaborations with local and national stakeholders. The Institute conducted much original research in-house and then helped various entities to implement them on an industrial level: be it in Jupiter's refinery in Pauillac, in the plant of Saint-Symphorien, or in the rocket testing sites of L.R.B.A. The Institute's long-term strategy, however, seemed somewhat ambiguous. On the one hand, it was designated as an internal laboratory of U.C.R., funded by and serving the local resin industry. On the other hand, the Institute was notoriously in conflict with the U.C.R., the same institution it was supposed to assist. The Institute competed with it for influence over the direction of the industry and tried to emancipate itself from its tutelage by engaging in the collaboration with big chemical companies or national institutions. Recall that in the Fréjacques Commission, there was not a single resin company, but its members came from French chemical giants, such as Saint-Gobain, Pechiney and Total.

Another remarkable feature of the period was the short-term application-focused projects that dominated the Institute's work. When a given resin-based product was thought to be more promising, total effort was put into it. This is in stark contrast with the interwar period, when all the topics were developed at the same time. The ambition to construct a more integrative and overarching approach towards resins, which defined the Institute in the interwar period, slowly faded away after 1945. The shift towards publications in more general reviews, such as *Bulletin de la Société chimique de France* instead of *Bulletin de l'Institut du Pin,* brought further the disintegration of the elements that had kept resin chemistry together as a discipline. The key observation to make here is that the Pine Institute moved from being the heart of resin chemistry to an institution that happened to work on resin-based compounds. Notably, harvesting and actual practices in the forest, which used to be at the heart of the resin chemistry project at its inception, ceased to be of interest to the Institute's researchers.

This evolution was the result of a number of factors: personal ambitions of the director, the rise of petrochemistry that set a benchmark for others to follow, and finally, the slow decline of the French resin industry itself. To fully understand the dynamics of the 1960s and the reasons that brought the Institute to the verge of closure in 1969, it is necessary to delve into one final question: the complex relationship between the Pine Institute and the local industry from a strictly economic and financial point of view. This is the main subject of Chapter 4.

Chapter 4

CAN SCIENCE FEED ON A CRISIS: THE RELATIONS BETWEEN THE PINE INSTITUTE AND ITS ECONOMIC ENVIRONMENT

> "*Dans notre siècle, la Science est l'âme de la prospérité des peuples.*"
> (In our age, Science is the soul of prosperity of nations)
> Louis Pasteur

> "*La Science a besoin d'argent.*"
> (Science needs money)
> Henri Poincaré

The modern innovation paradigm has two essential parts. The first was probably expressed most skilfully by the famous French chemist and pioneer of microbiology, Louis Pasteur (quoted above): "Science is the soul of prosperity of nations." [Pasteur, 1927] This idea has been emblematic of the way of thinking of entire generations of scientists for more than a century all over the world. Implicitly, the idea also became ubiquitous among politicians and the general public after the American engineer, inventor and science administrator, Vannevar Bush, delivered his famous speech "Science, the endless frontier" [Bush, 1945]. If science is so essential to economic and social progress, one may wonder how it can be cultivated and grown to secure the prosperity of future generations. A few decades after Pasteur, another great Frenchman, Henri Poincaré, provided an essential hint (also quoted above): "Science needs money." [Poincaré, 1904] It appears that a virtuous circle is formed: the more nations invest in science, the more science gives them back.

Nonetheless, the relationship between science and economic prosperity is never that simple; for example, times of economic hardship can stimulate research, as happened in the case of the British laboratories trying to save the traditional indigo blue industry in colonial India from the rise of the synthetic dye at the beginning of the 20th century [Kumar, 2012]. Conversely, a period of affluence and economic security can slow down knowledge diffusion and research, which happened, for example, to agricultural chemistry under the Second Empire in France (1852–1870) [Jas, 2000]. However, studies on historical cases that do not follow the traditional equation between economic prosperity and scientific progress remain few and far between. This chapter aims to fill this gap, to an extent, by presenting an example that casts a shadow of doubt on the traditional perspectives of the relations between science and prosperity. It shows that, at least under certain conditions, economic struggles may translate into substantial improvement of scientific funding.[143] More specifically, the chapter examines the funding patterns of the resin research units in the Landes over multiple decades and cross-references them with the overall economic results of the local industry.

1. Vèzes' Laboratory Funding (1900–1921)

To better understand the local economic context during the existence of the Pine Institute, the origins of the resin industry presented

[143] An early presentation of the results provided in this chapter was published in [Krasnodębski, 2017].

in Chapter 1 have to be reread through a new lens. As previously mentioned, in the last three decades of the nineteenth century, the region experienced a prolonged economic crisis. The huge price fluctuations led to social discontentment. This is because the gum was the property of the forest owners and not the forest workers who harvested it. Forest workers earned a small fixed sum for each litre of gum collected, as well as an annual salary after the sales were made. Consequently, they were the first to suffer if resin owners chose to stockpile the product and sell it later for a higher price. In practice, years of good fortune were followed by years of misery. In 1891, during massive and violent strikes against low salaries, nine workers were killed by the police. In 1907, another region-wide mobilisation occurred, resulting in many more deaths and a military intervention [Alquier, 2009, p. 6]. General Armand Jacquey, a deputy from the Landes, after having dispersed one of the illegal manifestations that year, called the right to strike "a monstrosity, a crime against the civilisation. This is the return to the Papuan regime". The comparison should not be surprising, given that only a few decades earlier, the Landes inhabitants had been compared to "the Hottentots" by some members of the French economic elite [Sargos, 1997, p. 56]. The French state was believed to have civilised the locals by transforming the wasteland into one giant resin factory.

The inhabitants of the region were acutely aware of the scale of the crisis. In 1901, the local journal *Petite Gironde* deplored the disorganisation of the local industry, and some commentators even expressed their disbelief in the possibility of recovery: "The situation is disastrous in the Landes; misery is growing, people are emigrating, and soon no more *résiniers* will be left." [cited in Barraud, 1931b, p. 3] It was in this political and social atmosphere, amidst another grave economic crisis, that the local authorities solicited the Faculty of Science for help. As explained in Chapter 1, scientists led by Maurice Vèzes answered the call.

The Laboratory of Chemistry Applied to the Resin Industry was established at the demand of the local authorities in the premises of the Faculty of Science of Bordeaux. As already mentioned, Vèzes underlined that its vocation was to help the local industry through

Table 10: Contributors to Vèzes' laboratory budget in 1903/1904 [Vèzes, 1905]

Source	Amount (frs)
Various communes in Landes and Gironde	1,990
Chambers of commerce of Bordeaux, Mont-de-Marsan and Bayonne	1,300
City of Bordeaux	1,000
Departmental councils of Gironde and Landes	1,000
Others (company, firms, associations)	445
Self-funding (service fees)	232.5
Total	5,967.5

applied research. To understand its position in the complex economic panorama of the early-twentieth-century Landes, the institutions and entities that contributed financially to its activities are examined in this section. The budget from the academic year 1903/1904 may serve as a starting point (Table 10).

Upon examining the budget, one immediately notices the small amount of private funding. The laboratory was to serve the industry, but these were, above all, public institutions of the local government that were involved in supporting its activities. The laboratory was one of the local government's tools to appease the citizens calling for immediate action against the crisis. Compared to stricter price or employment regulations that would transform wealth distribution patterns in the region, investing in science was deemed a much safer and less controversial solution for both forest owners and workers, among whom the faith in techno-scientific progress was still strong. The Bordeaux Faculty of Sciences, which hosted the laboratory in its buildings, should also be added to the list of contributors.

The information about the laboratory's budgets is derived from incomplete activity reports, but despite their gaps, they help to reconstitute a certain number of phenomena. First of all, in spite of the crisis, violent strikes, and social upheaval, the budgets remained relatively stable for the first seven years of the institute's existence (Figure 34).

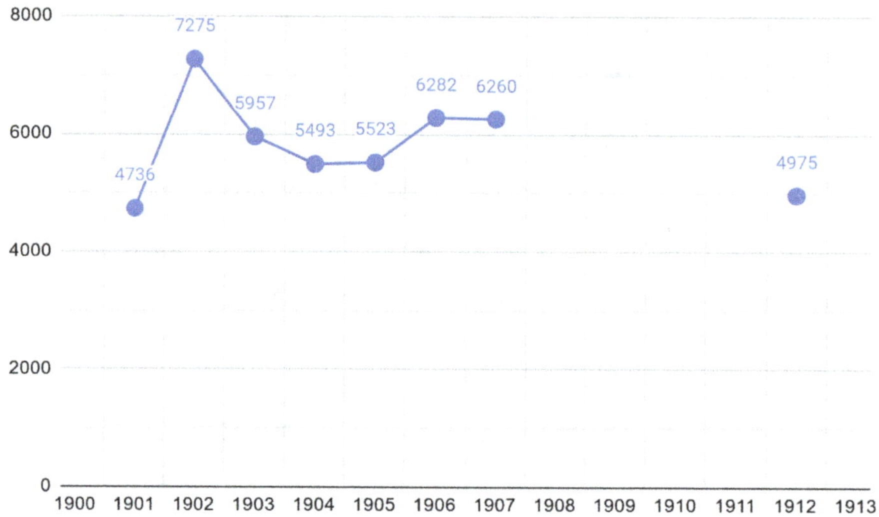

Figure 34: The resin laboratory's budgets in francs (1901–1912).[144]

The last available data point is for 1912, after a four-year gap. Adjusting for inflation, the budget for 1912 was equal to 4,308 francs from 1907. In other words, a substantial reduction in funding took place, around 32 per cent, between 1907 and 1912. Since there is no exact data for other years, it is not known whether this was an exceptionally bad year; however, it can be argued that it was part of a more pronounced tendency. The data concerning Vèzes' publications provided in Chapter 1 (high productivity in the 1900s and low productivity in the 1910s, see Figure 3) seem to confirm the hypothesis.

How can this sudden loss of funding be explained? In its first years, the laboratory was pretty successful in collecting donations from various entities. This appears to have changed after the resin crisis ended. With the end of strikes in 1907–1908, the forest workers managed to negotiate an improvement in working conditions and remunerations. Additionally, the situation in the global markets was also becoming increasingly favourable. Between 1908 and 1911,

[144] Histogram established from data in the activity reports of *Laboratoire de chimie appliquée à l'industrie des résines* from years 1901 to 1908, 1912.

turpentine spirit prices doubled [Maydieu, 1931]. But it was, in particular, the shortage of raw materials triggered by World War I that massively increased the consumption of resinous products and, as a consequence, their prices. "A new era of prosperity took place during the War of 1914–1918 and especially after the war, at the moment when the resin prices increased alongside the exchange rates," explained Marcelle Barraud [1931b, p. 1]. The estimates of American economists confirm these general tendencies concerning the French resin industry [Woolsey, 1921].

This story reveals an intriguing paradox. The laboratory of Maurice Vèzes was intended to combat the economic crisis, but in the 1910s, social unrest ended, and prices went up. The crisis was over, and the laboratory lost its *raison d'être*. There was no longer any incentive to invest in science because the business paid well, and local communities were less interested in contributing to research and development, which was perceived as irrelevant to their wealth. As a consequence, the laboratory's budget shrank, and Vèzes did not have enough funding to conduct his own research.

2. The Pine Institute and the (New?) Crisis of 1920–1940

This curious inverse relationship between the prosperity of the region and its scientific institutions became even more pronounced in the interwar period. As already explained, in the 1920s and 1930s, interest in resin chemistry exploded, and the Pine Institute became the international hub of this new flourishing discipline. And yet, even though the Institute boasted over its collaboration with the industry, the actual impact of this collaboration was far more ambiguous.

Change in Budgets and Resin Prices: An Inverse Relationship

After Georges Dupont's call for the creation of the Pine Institute in 1920, and thanks to him advertising in the local newspapers, more and more organisations and entrepreneurs started regular payments

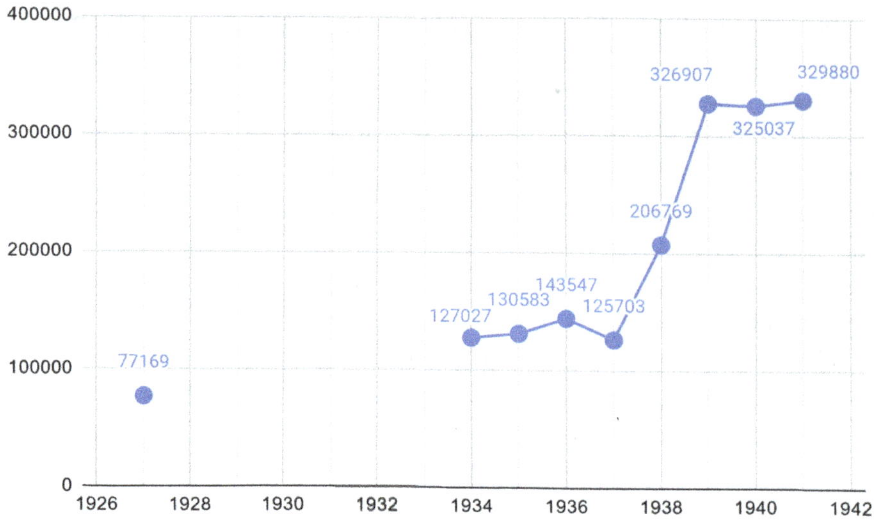

Figure 35: Pine Institute's budgets in francs, adjusted for inflation (1927 prices).[145]

to cover the new Institute's fees. Full data on budgets available throughout the interwar period are not available, but the numbers that exist indicate rapid growth (Figure 35).

First, it is worth noting that the budget almost doubled between 1927 and 1934. This rise is even more pronounced if we recall that many of its researchers were remunerated by external agencies (such as the Fraud Repression Service). Second, the jump between 1938 and 1939 was most likely due to the transformation of the laboratory into a "faculty institute", meaning that some contributions from the Pine Institute Association were from this moment included directly in the new Institute's budget. All of this to say that the increase in the budgets was most likely more gradual as some funds were not revealed before 1938 in the laboratory's reports for accounting reasons.

To understand the reasons for these augmentations, the sources of the Institute's funding in 1927 and 1938 are compared in Figures 35 and 36.

[145] Established from the activity reports published in the pages of the *Bulletin de l'Institut du Pin*.

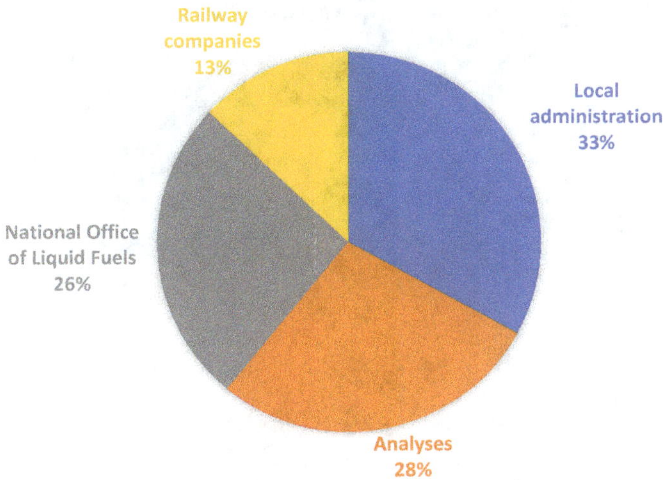

Figure 36: Source of the Institute's funding in 1927.

The 1927 budget has many particularities as compared to the one from 1903, as shown in Table 9. Most visibly, more than a quarter of the overall budget comes from self-funding, namely the analyses of the resinous products on behalf of different businesses. Overall, however, the local government institutions (meaning both departments and communes) provided one-third of the budget, making them the most important stakeholders of the Pine Institute but much less important as compared to 1903, when they had been exclusive sponsors of the resin laboratory. Almost 26 per cent of the funding came from the national government through the National Office of Liquid Fuels. This state agency was actively looking for various alternatives to petroleum and funded studies, among others, on the gasification of wood. Finally, the last category of stakeholders comprised the private railway companies interested in the properties of pine wood in railway construction. The railways and the National Office for Liquid Fuels accounted for almost 40 per cent of the overall budget, which is important considering that these institutions were interested in studies on pine wood, not pine resin. It seems as if in 1927, the Pine Institute was struggling to get the resin industry behind its back and tried to extend its expertise on the previously

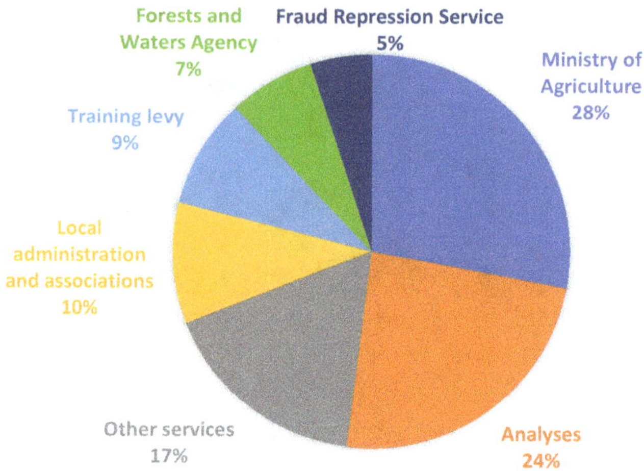

Figure 37: Source of the Institute's funding in 1938.

uncharted terrain of pine wood use in different branches of the industry. And yet, the 1938 budget shows different tendencies (Figure 37).

In 1938, the government agencies (Ministry of Agriculture, Forest and Water Administration, and Fraud Repression Service) funded no less than 40% of the Institute's overall budget. The Ministry's subvention was attributed through a number of grants coming from the so-called *Comité central des produits résineux* (Central Committee of the Resinous Products), whose mission was to help the region combat the consequences of the economic crisis. Thus, while the state remained the key stakeholder, it explicitly directed its funding to research on resin chemistry, unlike in 1927. Another noteworthy feature of the 1938's budget was that self-funding continued to play an important role, even though the role of the local industry was more prominent. Notably, in the category "other services", the income from the sales of the Pine Institute Bulletin, as well as membership fees of the Pine Institute Association and similar sources, is included. They had not been included in the earlier reports due to a different organisational structure of the Institute. A more explicit contribution of the local industry was the training levy (*taxe d'apprentissage*), which, to put it simply, was a tax-deductible contribution whereby businesses could choose a training or educational institution and fund its activities. Many local companies chose to support the Institute, and Georges Brus regularly encouraged them to do so with personalised letters. Finally, what is surprising is that, by 1938, the overall contributions of the local government institutions were minuscule compared to 1927. To sum up, between 1927 and 1938, three tendencies are clearly identifiable: (1) the growing role of the national government and, to a lesser extent, businesses, (2) the consistent level of self-funding, and (3) the diminishing role of the local government.

To provide a complete understanding of the Pine Institute's status in 1938, it should be mentioned that 58 per cent of its expenses were allocated to staff salaries and the remainder on equipment. That year, five out of nine employees were paid directly from the Institute's budget, while the others were employed by the Fraud Repression Service and the Bordeaux Faculty of Sciences.

Overall, the situation of the Institute appears to have continuously improved throughout the interwar period. Was it due to its successes in solving the industry's problems? The traditional belief regarding innovation is that applied science leads to economic prosperity, resulting in more funding for scientific research and creating a virtuous circle of progress and prosperity. However, examining the evolution of the prices of resinous products in France during the same period suggests a different interpretation. It seems that the Pine Institute's success was positively correlated with the economic difficulties of the region, as illustrated in Figure 38.

Figure 38 reveals many interesting features. First, there was an abrupt drop in prices in 1921. This fall coincided with the establishment of the Pine Institute at the beginning of 1922. It seems that after years of economic growth, paralleled with the decline of the old resin laboratory, 1921 was a sobering shock to the resin industry. Its members realised that their well-being was not unconditional. In that year, Dupont and Vèzes managed to persuade a sufficient number of stakeholders that the Pine Institute was necessary to avoid future crises, even though this was a mere rebranding of the existing

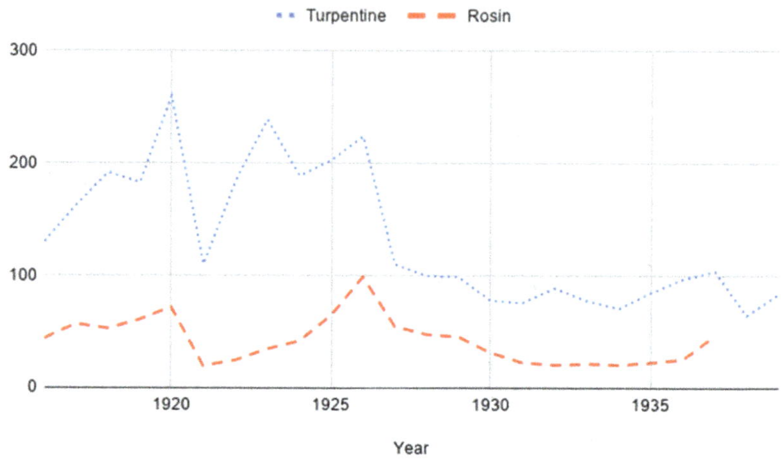

Figure 38: Prices of resinous products (frs/100 kg), adjusted for inflation (1916 prices).[146]

resin chemistry laboratory. They "sold" their expectations of future research results. However, the 1921 crisis was short-lived. The economic situation was considerably better in 1922, and it looked as if the local investors lost interest in science in the first half of the 1920s. Of course, Dupont doubled his efforts to energise their involvement. He noted that returning to pre-World War I prices would be "a terrible disaster" for the region [Dupont, 1926b, p. 413]. He stressed that the Pine Institute was "the necessary link between science and the industry, a link still insufficient without doubt, to answer all the needs", and he urged the local businesses and authorities to secure additional funding. His pleas fell on deaf ears. As noted, in 1927 (the budget being based on 1926 conditions), 40 per cent of the Pine Institute's funds came from the public institutions that were not connected to the local resin industry. Without them, the future of the Institute would have been called into question. In other words, the Pine Institute would not have been able to function without external investors, since the resin-producing

[146] Compiled from data in *Bois et Résineux*, years 1927–1939.

companies and local communities were not convinced of the necessity of research in the field of resin chemistry.

However, everything changed in 1927 when the most severe economic crisis struck the resin industry. The fall of prices had most likely two major reasons. Firstly, the monetary reform by Prime Minister Raymond Poincaré aimed to stabilise the currency, which led to turpentine losing 50 per cent of its value. Secondly, the resin market slump in the United States (US) reduced French turpentine's prices by 33 per cent.[147] It was the overproduction of American pine gum, which continued to be cited as the main reason for the troubles in Aquitaine in the long term [N. N., 1928b]. As France exported around half of its production abroad over the period, it was very sensitive to changes in the international markets. Finally, the poor organisation and lack of coordination between local producers exacerbated the crisis [N. N., 1927d]. The key point is that the resin crisis preceded the Great Depression, which further ravaged the trade. The 1927 crisis led, as in the 1890s, to social unrest. Forest workers felt again that they were the only ones suffering from the consequences of the crisis. Labour unions became more active and organised large-scale strikes in the region.

Some forest owners failed to comprehend the claims of the workers. In the pages of *Bois et Résineux*, one of them notably wrote about forest workers in the following way: "I consider them much happier than labourers in the city ... They can raise one or a few cows." He added: "I think that the great fault of our times is to confuse money with happiness." [N. N., 1930b] His arguments did not find favour with the local population, and the unionisation proceeded further [N. N., 1931b]. Some thought the Pine Institute Association could potentially act as a mediator or an overarching organisation for all the feuding parties, but the project failed [Groc, 1931]. Overall, it was this tumultuous period that later led to the establishment of the *Union Corporative des Résineux*, which is introduced in Chapter 3.

[147] This was the contemporaneous perspective [N. N., 1927d]; Historians today present a more nuanced interpretation. See [Blancheton, 1997].

Selling Hopes: Scientific Promises as Products of the Resin Laboratory

From the Institute's point of view, the 1927 crisis was the beginning of a new era. In 1930, an article in *Bois et Résineux* titled "Resin crisis: first remedies" proposed three policies to mitigate the effects of the economic downturn: combating counterfeit resinous products, using turpentine in public infrastructure, and incorporating rosins into soaps. However, the most important and overarching policy advocated in the article was to grant an important subsidy to the Pine Institute. The Institute's scientists were actively involved in promoting their activities by arguing that prosperity would come back to the region thanks to science. Barraud asserted, for example, that the Landes Forest constituted "considerable wealth" and that "the progress of science will enable it to be exploited rationally, to conserve it, and live from it" [Barraud, 1931b, p. 3].

Dupont's rhetoric was even more emphatic. He explained in 1931 that an end to the resin industry would be "disastrous" for the entire region as, according to him, the Landes Forest would simply disappear if not carefully maintained by forest workers — "Giving up [on the resin industry] would mean returning to the ancient state of the land: one million hectares of pestilent swamps and moving dunes, and Bordeaux, the region of great wines, would be threatened by the invasion of sand once more." [Dupont, 1931] According to Dupont, investing in the Pine Institute was not only necessary in order to sustain 20,000 families living on the resin industry but also to preserve the entire region, if not the whole nation, from imminent economic and environmental calamity. Dupont invoked in his speeches the old fear about the threat of moving dunes and sandstorms that could ravage the land if it was not for a carefully managed forest.

Another line of argumentation of the Institute's stakeholders was to insist on the inadequacy of the Institute's financial means as compared with other nations. In 1931, one of the leaders of the Pine Institute Association lamented the German superiority in chemistry and cautioned his fellow producers that the Germans were better prepared to produce new products from turpentine than the French. As a consequence, he explained, a chemical synthesis industry must be established in France with the help of the Pine Institute: "We should make this laboratory thrive, to provide it with the resources necessary to make it possible to manufacture the largest number of possible derivatives from turpentine and rosin in our region." [Maydieu, 1931b] The scientific work of the Pine Institute was presented as the sole way to find new market outlets and salvage the resin industry. In 1932, Barraud warned that "foreign countries are, in fact, much better equipped than the Landes" [Barraud, 1932, p. 157]. In 1935, Brus wrote that it was sufficient to read "American and German scientific journals" to see "our lag both in terms of technical questions and questions of pure science" [Brus, 1935, p. 195].

This carefully elaborated narrative proved to be very effective. In 1934, *Bois et Résineux* announced in strong terms: "The future of resin products is in science," and asserted that to overcome the crisis, "it is necessary to form men of science, experienced and knowledgeable technicians who are farseeing and courageous ... to provide this intellectual elite, which will be entrusted with the mission of defending the Forest of Landes ... with fully equipped scientific laboratories." [N. N. 1934] How would they do that? An article "To fight the crisis — *taxe d'apprentissage* to the Pine Institute" in *Bois et Résineux*, which called on local companies to pay the already mentioned training levy in favour of the Institute, left no doubt. Even Victor Grignard, a Nobel Prize winner in chemistry, supported the promotion of the Pine Institute: "If we do not subsidise our laboratories, it will soon be the end of the French science ... The French resin industry has its laboratories: the ones in the Pine Institute, created at the Faculty of Sciences; they live modestly, from insufficient subsidies, smaller every year." [cited in Brus, 1935, p. 199]

While the Institute's promotional activities led to substantial growth of its budget, the crisis continued to ravage the region with no sign of improvement in the first half of the 1930s. Voices of despair regularly appeared in the local press: "It has been forty years that I've harvested resin; never have I seen a similar catastrophe," wrote one forest worker [N. N., 1934]. "Today a forest worker cannot live anymore, the trade is dying, the Landes cry out in distress," added a local businessman [Merklen, 1934, p. 227]. The low prices of gum resins led to a steady decline in production that, on its own, led to a growing number of people abandoning forest jobs and emigrating from the region, resulting in a decrease in the amount of resin produced.

The voices of local inhabitants eventually reached Paris, prompting the French parliament to vote on a law for the reorganisation of the resin products market on 2 July 1935. Later the same year, a special fund for compensating the resin workers and owners was established. This fund would later become the Central Committee of the Resinous Products of the Ministry of Agriculture. Brus personally intervened to persuade the Committee to expand its activities to research funding. The Committee allocated in 1937 (for 1938) no less than 100,000 francs to support the Pine Institute, and it remained one of its leading funding sources in the following years.[148] It is worth mentioning that the directors of the Pine Institute did not solely concentrate on promoting science and its potential contributions to the local industry. They also actively advertised the resin industry to potential business partners who might not have been acquainted with pine resins and their diverse applications. In 1938, Brus even ordered a promotional poster that was to showcase the variety of products that could be extracted from maritime pine with the help of scientific research.

The poster displayed in Figure 39 illustrates the diversity of final products that could be extracted from pine resin and wood: fuels, solvents, printing inks, etc. There were around 30 categories, many of them corresponding to the Pine Institute's research themes. However, these included all possible and projected uses. In practice, the consuming markets were not as diverse. Figures 40 and 41 show the consumption of turpentine and rosin in 1937.

What immediately stands out concerning turpentine consumption is that 87 per cent of it was used as solvent (in paints, varnishes,

[148] Departmental Archives of Gironde, Pine Institute Archives, folder "Correspondance 1937 G.B., C". Letter to senator Caillier, 31 May 1937.

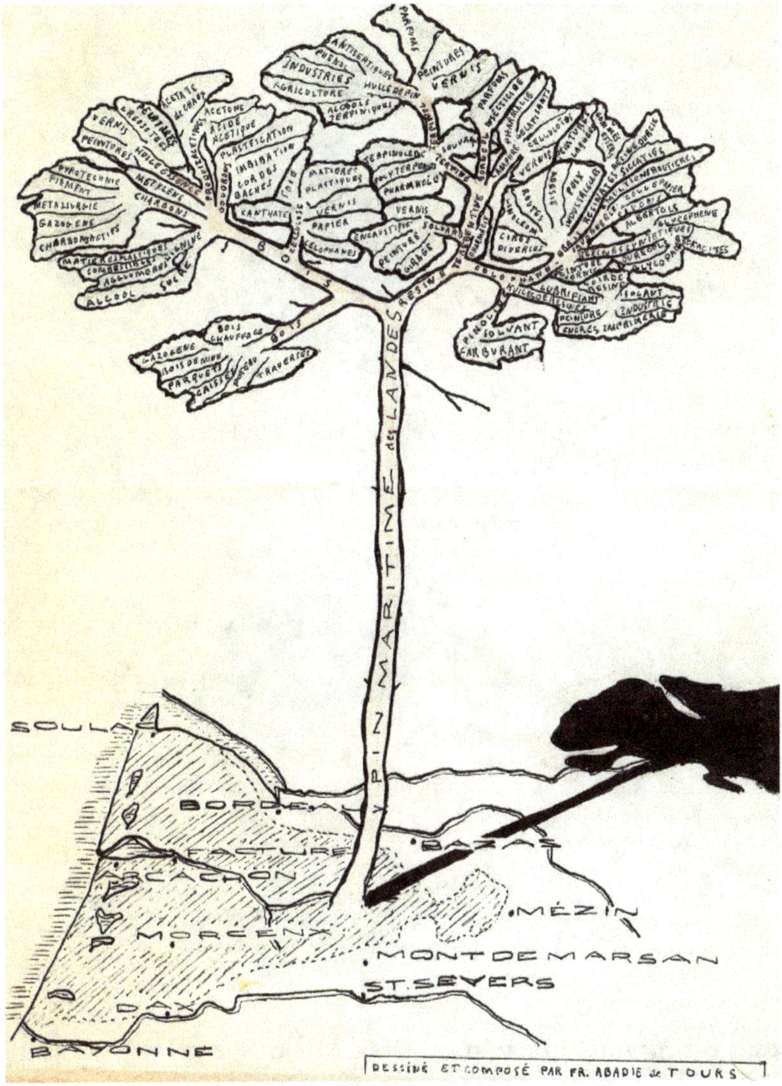

Figure 39: Poster promoting the resin industry "Pin des Landes" (1938).
Source: Departmental Archives of Gironde, Pine Institute Archives

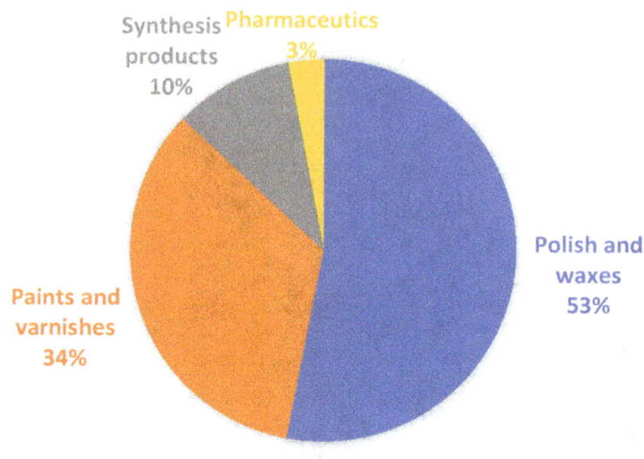

Figure 40: Turpentine-consuming markets in France (1937).[149]

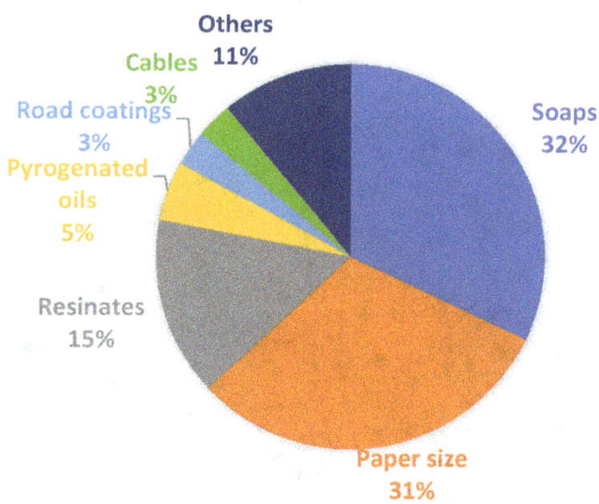

Figure 41: Rosin-consuming markets in France (1937).[150]

waxes, polish, and similar products). Various synthesis projects on which the Institute worked on throughout the 1920s and 1930s corresponded to barely 10 per cent of the overall consumption. This explains why the Institute was reluctant to work on turpentine after World War II; in spite of its high prices, turpentine was valued on traditional markets and failed to gain new ones (in France).

[149] Established from data in [Sargos, 1938].
[150] Established from data in [Sargos, 1938].

The consuming markets for rosin were more diverse, and the Institute's contributions more visible in the development of road coatings, pyrogenated oils, and resinates. It also occasionally studied the use of rosins in the paper industry. At the same time, the paper mills rarely called upon the Institute's expertise in the 1930s, and it is necessary to reiterate that local companies also actively contributed to the development of industrial processes for the manufacture of rosin derivatives. While the Pine Institute certainly played an important role, its place in the establishment of new markets for rosin should not be overestimated. If the Institute had not existed, the industry might have lost a few smaller markets, but the overall manufacture and consumption trends would not have been radically altered.

And yet, despite the low resin prices, the decreasing production, and an overall modest impact of the Institute's research projects on the big trends in the resin industry, not only did the Institute's budget triple between 1927 and 1938, but its golden age was about to begin.

3. The Golden Age of the Pine Institute (1941–1960)

As explained in the previous chapter, the collaboration between the Pine Institute and the *Union Corporative des Résineux* (U.C.R.) was tense during and after World War II due to the Institute's close ties to the petroleum company, Jupiter. However, despite how strained relations with the U.C.R.'s directorship might have been, the organisation was, above all, a federation of different social groups that generally trusted the Pine Institute and its expertise and held its scientists in high regard. The U.C.R.'s (and then the N.U.C.R.'s) official policy was to support scientific research. Hence, its members contributed generously to the Institute's budgets. Data for the 1940s show the importance of the U.C.R.'s subvention in the Institute's income (Figure 42).

From its inception in 1900 until the mid-1940s, the resin laboratory and the Pine Institute relied on contributions from various entities. However, it appears that in the post-war period, the lion's share of the Institute's budget came from a single institution

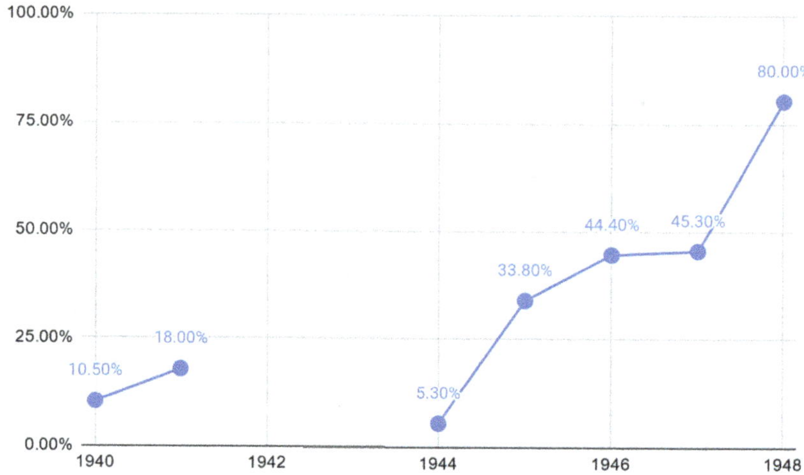

Figure 42: U.C.R.'s subvention share in the Institute's budget.[151]

representing the industry, namely the U.C.R (and later N.U.C.R). The business world was involved in research funding more than ever before. It is also remarkable to note the jump in funding in 1948. As will be explained shortly, it can be partly explained by a governmental subvention; however, it is likely that the U.C.R. itself decided to increase its contributions after the visit of its delegation to the US in 1947. During the trip, regardless of whatever conflicts might have taken place there between the Institute's and U.C.R.'s delegates, the French industry realised the importance of investments in science in the American resin industry and the size of its laboratories. The U.C.R.'s director wrote in October 1947: "[T]he brains of the French researchers are not inferior to the foreign brains, it is only a question ... of planning and funding." [Prat, 1947b]

Of course, this does not mean that the Institute was not seeking other funding opportunities and did not benefit from additional resources not visible in is budget outlines; for example, in 1941, the municipal council of Bordeaux decided to construct new buildings

[151] Established from data in: Departmental Archives of Gironde, Pine Institute Archives, box "Divers". G. Brus, "Le Fonctionnement et les Travaux de l'Institut du Pin en 1941", 1943; Also, data in *Bois et Résineux*, 1940–1950.

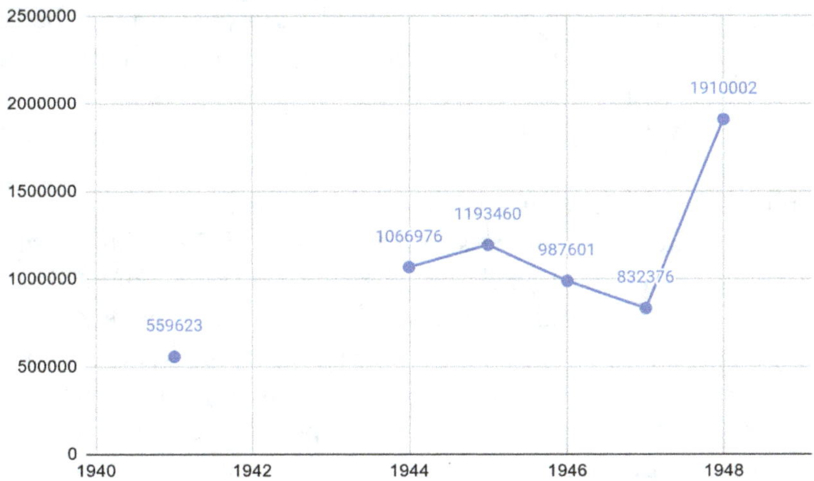

Figure 43: The Institute's budgets from 1941 to 1948, adjusted for inflation (prices 1941).[152]

specifically for the Institute, and from 1940, some of its employees were paid directly by the young CNRS.[153] Overall, the immediate postwar period was an age of prosperity for the Institute (Figure 43).

In 1948, the Institute received an important subvention from the Ministry of Agriculture and from the CNRS to acquire modern equipment (the subvention was most likely passed to the Institute via the U.C.R., which explains the spike) [N. N., 1949b]. Brus was grateful for the contribution but noted "the state of [the Institute's] equipment was much better than of its employees" and complained about the low salaries in his institution. In *Bois et Résineux*, Brus explained: "Unfortunately, the subvention is insufficient for the Institute to orient itself more actively to pure research." [N. N., 1947] It should be reiterated that Brus said this after the most prosperous year in the Institute's history. In 1948, the wide network of

[152] Established from data in: Departmental Archives of Gironde, Pine Institute Archives, box "Divers". G. Brus, "Le Fonctionnement et les Travaux de l'Institut du Pin en 1941", 1943; Also, data in *Bois et Résineux*, 1940–1950.

[153] Departmental Archives of Gironde, Pine Institute Archives, box "Divers". G. Brus, "Le Fonctionnement et les Travaux de l'Institut du Pin en 1941", 1943.

the Pine Institute's collaborators totalled 42 people, and its budget was as high as 16 million francs. For comparison, in Grenoble, the laboratory of the future Nobel Prize winner, Louis Néel, had a budget of only 4.5 million francs (without salaries) and six scientists. Another emblematic laboratory from the period, the Centre of Chemical Metallurgy in Vitry-sur-Seine, directed by Georges Chaudron, had 20 employees and a budget of 7.6 million francs (also without salaries) [Teissier, 2014, p. 116]. Comparing the public laboratories' budgets was tricky since many expenses were covered by the respective universities, especially concerning employment. However, it is clear that the Pine Institute was far from lagging in terms of funding. Even if around 60 per cent of its budget was spent directly on remunerations, mirroring the proportions from before World War II, it still did not fare worse than some of the most prestigious French laboratories in the late 1940s.

The archives do not provide the full picture of the budgets of the Pine Institute in the post-war period; however, the U.C.R./N.U.C.R. subventions highlight some developments of the budget, as shown in Figure 44.

After a prosperous period between 1948 and 1953, the N.U.C.R.'s contributions to the Pine Institute decreased, but even then, the subventions remained more than twice as high as before 1948. In the early 1950s, the non-N.U.C.R. funding amounted to 5 or 6 million francs per year (around 25 per cent of the N.U.C.R.'s funding).[154] In 1956, for example, the budgetary situation improved due to a single 5-million-franc grant from the local paper industry to fund research on cellulose. In total, the Institute's income that year reached 21 million francs, which, when adjusted for inflation, was around 33 per cent smaller compared to the budget in 1948. Interestingly, the Institute's expenses were higher and amounted to 25 million francs, but the surplus from the previous year made it possible to cover the gap. In fact, by 1956, the Institute set aside almost 22 million francs, which was more than one year's income. In other words, it was

[154] Departmental Archives of Gironde, Pine Institute Archives, file "U.C.R.". G. Brus, "Rapport sur le fonctionnement et les travaux de l'Institut du Pin", 1956.

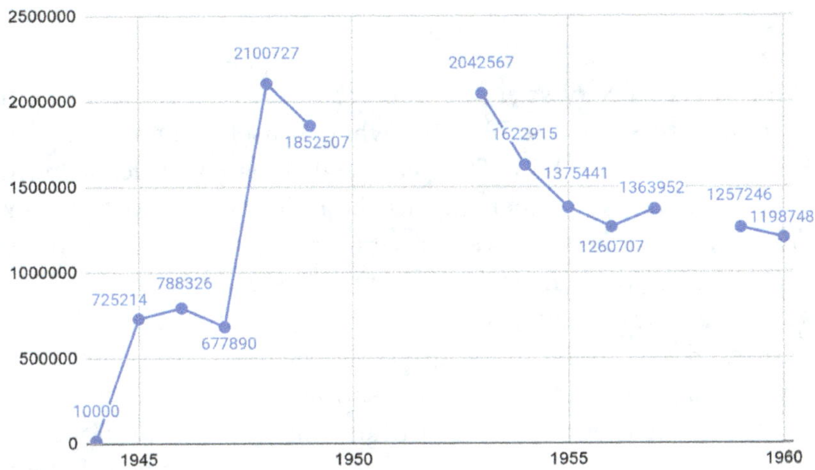

Figure 44: U.C.R.'s (N.U.C.R. after 1948) subvention to the Pine Institute, adjusted for inflation (1944 prices).[155]

operating with important margins that could permit it to cover losses for a few years. This strategy was perhaps necessary considering the instability of yearly subventions. Brus regularly complained that he could not offer his employees a competitive salary. In 1956, the Institute spent around 16 million francs on salaries alone, and Brus requested additional funding from the N.U.C.R. He notably underlined that there were but five chemical engineers working full-time on resin chemistry, properly speaking (so excluding analyses, biology, paper, etc.), compared to eight in 1948. In practice, the situation was even worse, as some chemists had to work half the time on analyses of samples and not on original research, strictly speaking.

Brus continued to promote the Institute and the role of science in the reconstruction of the resin industry throughout the 1940s and the 1950s. In 1949, he wrote that "Scientific research always pays and ... the statement 'Knowledge is Power' ... is not an empty slogan" [cited in N. N., 1949b]. In 1953, the Institute had its own stand at the *Salon*

[155] Established from data in *Bois et Résineux* and the activity reports of the Pine Institute (1953–1957) in the Departmental Archives of Gironde, Pine Institute Archives, box "U.C.R.".

de la Chimie (Salon of Chemistry) that took place in Paris, where it showcased a range of products extracted from pine gum and emphasised the importance of scientific research in the reconstruction of the resin industry [N. N., 1953]. But while Brus fought for scientific research, especially in the 1950s, the social climate was deteriorating. On the one hand, some members of the industry suggested the Pine Institute's research results were disappointing and that its funding should be reduced.[156] In 1957, a certain Lescarret wrote in *Bois et Résineux* that the only way to save the resin industry was to find new uses for gum and its products. However, he said, "I do not believe that the small laboratories in the factories that manufacture resin products ... not even the Pine Institute can give us any hope. Not because the scientific quality of the Institute's directors is questioned, but because their means are insufficient to arrive at any new result fast." [Lescarret, 1957] Although Lescarret admitted funding was inadequate, his scepticism illustrated a broader disenchantment persisting in the industry that accused the Pine Institute of neglecting applied science. If research institutions were not able to redeem the resin industry, why fund them at all? Brus had to constantly respond to this type of criticism and underline in a somewhat exaggerated manner the importance of the Institute's research:

> [O]ne should not believe that [the Institute's research] is excessively theoretical because its applications do not appear immediately. The evolution of the entire industry depends on research, and the resin industry is not an exception. I remind you of it, as I do not ignore the fact that some of you [members of the resin industry] find the Pine Institute overly focused on theory. It must be theoretical. Be reassured that it is in your best interest.[157]

On the other hand, the self-perception of the industry's state from an economic standpoint was increasingly negative. A glance at

---

[156] Departmental Pine Institute Archives, Pine Institute Archives, file "U.C.R.". G. Brus, "Rapport sur le fonctionnement et les travaux de l'Institut du Pin", 1956.

[157] Departmental Pine Institute Archives, Pine Institute Archives, file "U.C.R.". G. Brus, "Travaux récents effectués à l'Institut du Pin", April 1950.

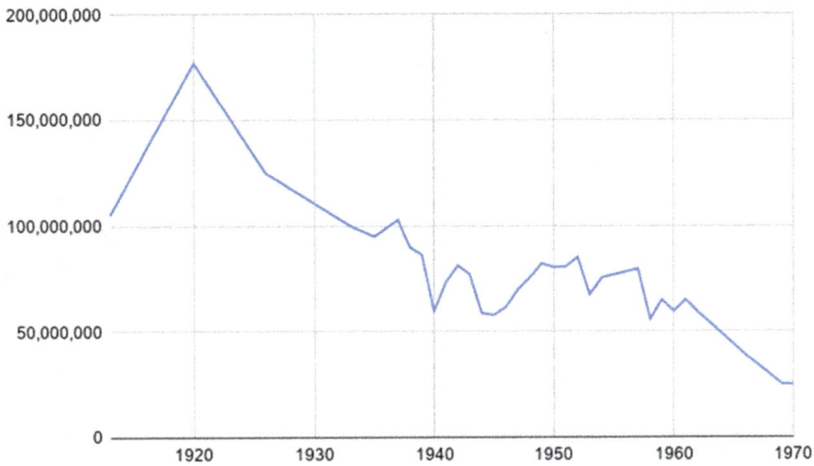

Figure 45: Gum production in France in litres (1910–1970).[158]

a selection of titles in *Bois et Résineux* from the period suffices to demonstrate this trend: "Why is the Landes Forest region in peril?", "The crisis of the Landes Forest", "The crisis of the resin products market", "Perspectives on crisis, perspectives on salvation", "End of crisis? Or the light of hope", and "The necessary jump or the last chance".[159] This last article from 1956 affirmed that "[t]he gum production is in danger, we would even say, in mortal danger". In other words, the prevailing narrative was as sombre as in the 1930s. As for gum production in France (Figure 45), it continued to fall between 1920 and 1970, with the prices varying widely but also slowly declining (Figure 46).

Even though the price of resinous products remained, on average, similar between 1926 and 1970 (with a massive slump between 1927 and 1935 and a mid-1940s spike), the production was systematically decreasing. Between 1926 and 1949, it fell by 33 per cent. Over the same period, the Institute's budget multiplied a few times. Interestingly,

[158] Established from data in *Bois et Résineux*, 1926–1969.

[159] *Bois et Résineux*, 10 March 1949; *Bois et Résineux*, 10 December 1949; *Bois et Résineux*, 30 November 1952 and 30 July 1954; *Bois et Résineux*, 10 February 1953; *Bois et Résineux*, 10 September 1953; *Bois et Résineux*, 20 March 1956.

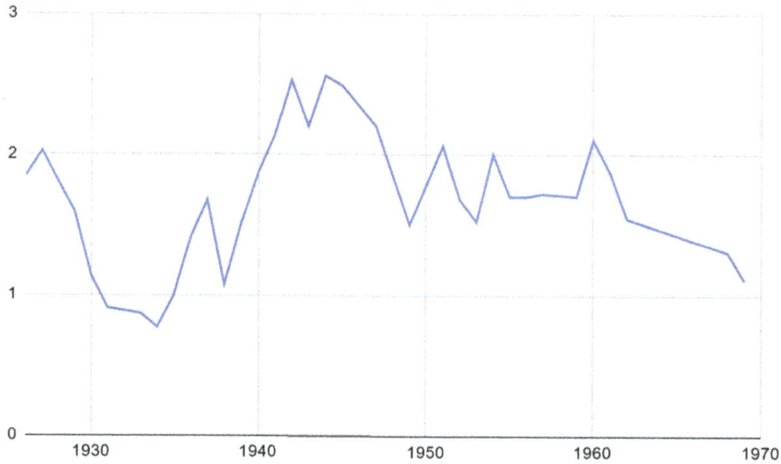

Figure 46: Price of gum (1910–1970, francs/litre), adjusted for inflation (1926 prices).[160]

the most prosperous year (1948) in the Institute's history followed a period of very high prices but also of very low production. The 1950s, when the budgets were shrinking and Brus had to double his efforts to convince the industry that investing in research was necessary, were a period of relative economic stability.

4. Prosperous Death: The End of the Pine Institute's Resin Research

In 1961, René Rusterholz, deputy director of the N.U.C.R., reassured local stakeholders that measures had been taken to address the difficulties faced by the region. He emphasised the creation of a new "applied research" laboratory to exploit the already known but unused technologies. This new laboratory, established in 1959 by Roger Lalande, had a peculiar status: it was housed in the same building as the Pine Institute, and its researchers indicated the Pine Institute as their institutional affiliation. However, in reality, all of its researchers were paid by the Faculty of Sciences of Bordeaux and the Bordeaux

[160] Established from data in *Bois et Résineux*, 1926–1969.

School of Industrial Chemistry, and it had separate budgets from the Pine Institute [Krasnodębski, 2016]. Brus cited, as the inspiration for its establishment, the Laboratory of Chemistry Applied to the Resin Industry of Vèzes to connect with the local resin history.

A new "technical service" at the N.U.C.R. was also established under the guidance of the Pine Institute, with a staff consisting of skilled technicians whose mission was to visit plants in the region in order to improve the quality of the products and processes.[161]

Rusterholz tried to convince local stakeholders that the N.U.C.R. was doing all it could to improve the situation because, in the 1960s, both the prices and production were deteriorating year after year. Not only were the narratives and expectations increasingly pessimistic, but economic indicators suggested that the future of the resin industry was in peril as well. In contrast, the Pine Institute was thriving. In 1961, after the 33rd Industrial Chemistry Congress, one participant said: "The eminent role played by the Pine Institute led many participants (especially the delegates from Spain, Greece, and Portugal) to express the wish that the Institute become a European Institute." [N. N., 1962]

An interesting shift in perspective was underway. Still, in March 1962, Pierre Legendre said that, in principle, he did not object to assisting foreign companies, but the priority was to safeguard the interests of the French manufacturers.[162] But already in July 1962, France, Spain, Portugal, and Greece signed an agreement to create an international trust of resin producers in the framework permitted by the Treaty of Rome (establishing the European Economic Community). "If the trust is established, the Pine Institute will be called to largely extend the scope of its activities," an internal report of the Institute stated in 1965.[163] It is possible that the Institute's

[161] Departmental Archives of Gironde, Pine Institute Archives, file "Lettres Techniques 1960–1961, N". Correspondence between the Pine Institute, N.U.C.R. and the company Sheby. First letter dated from 13 May 1961.

[162] Departmental Archives of Gironde, Pine Institute Archives, folder "Lettres Techniques 1962–1965, D". Letter from the Pine Institute to D.R.T. from 22 March 1962.

[163] Departmental Archives of Gironde, Pine Institute Archives, file "Institut du Pin, Ve Plan". "Recherche universitaire hautement spécialisée, construction et équipement de l'Institut du Pin", 25 February 1965.

leaders realised that the national industry was beyond saving and the only logical step for the Institute's future would be to connect with all European stakeholders. The same year, in 1965, Brus deplored the difficult financial situation of the Institute, especially in comparison with the American Olustee Naval Stores Research Station in Florida.[164]

Despite the challenging economic conditions facing the resin industry, the Pine Institute's situation was stable. In 1964, it obtained a single grant of one million new francs (there was a devaluation in 1958), which was equivalent to approximatively 32 million francs from 1948, from the French fourth economic plan for 1962–1965, aimed at rescuing the resin industry. In 1966, the Institute's budget reached 608,150 new francs, which corresponded to around 18.5 million francs in 1948. In other words, between 1948 (when the budget was 16 million francs) and 1964, the budget grew by about 13 per cent, even though the resin industry was in distress.[165] This rise can be partly explained by new openness towards the paper industry, but nevertheless, in 1966, the main field of study of the Institute was still the chemistry of resin products. The Institute received 80,000 francs directly from the Resin Compensation Fund, established in 1963 at the Ministry of Agriculture after the N.U.C.R.'s lobbying [N. N., 1963c]. In terms of staffing, the Pine Institute employed, in 1966, nine chemical engineers, fourteen technicians, and four administrative staff, not to mention the considerable number of employees remunerated by the Bordeaux Faculty of Science or the School of Industrial Chemistry of Bordeaux (the former School of Chemistry Applied to Industry and Agriculture).[166]

[164] Departmental Archives of Gironde, Pine Institute Archives, file "Institut du Pin, Ve Plan". "Recherche universitaire hautement spécialisée, construction et équipement de l'Institut du Pin", 25 February 1965.

[165] Departmental Archives of Gironde, Pine Institute Archives, file "Institut du Pin, Ve Plan". "Recherche universitaire hautement spécialisée, construction et équipement de l'Institut du Pin", 25 February 1965.

[166] Departmental Archives of Gironde, Pine Institute Archives, file "Institut du Pin, Ve Plan". "Recherche universitaire hautement spécialisée, construction et équipement de l'Institut du Pin", 25 February 1965.

However, budgets were not the only thing that improved in the 1960s. In 1960, the new campus of the University of Bordeaux was constructed. The Pine Institute could leave its old building at rue Paul Bert, in Bordeaux's city centre, and move into the new spacious facilities of the Faculty of Chemistry in the new campus (*domaine universitaire de Talence Pessac Gradignan*). Brus was not satisfied — the 860 m^2 occupied by the Institute in 1960 was insufficient for its numerous collaborators. His ambitions were much bigger. Considering the important grant in the framework of the fourth economic plan that concluded in 1965, Brus took a chance one more time and applied for more funds in the fifth economic plan launched by President Charles de Gaulle. He applied for no less than 8,500,000 new francs (over 16 times more than the budget from the record year 1948), which would serve for the construction of a new building of 6,000 m^2 — enough space for semi-industrial equipment, as well as for modern laboratory instruments, such as a mass spectrometer. Failing to receive the requested grant, Brus scaled down the project to 5 million francs, and it was accepted by the national government.[167] In 1970, thanks to this additional funding, the Institute moved into a new modern building comprising 4,000 m^2 of space, with 800 m^2 exclusively for semi-industrial installations.

While the Institute's budget multiplied and it participated in more ambitious projects (new buildings, semi-industrial equipment, international collaboration), the resin industry was on the brink of collapse. Prior to World War II, 100,000 m^3 of gum was manufactured in the Landes. The figure was around 38,550 m^3 in 1966, 25,250 m^3 in 1969, and 21,000 m^3 in 1970. This decline led to a dramatic decrease in the population living in the forest region. In 1945, there were 19,000 forest workers (*résiniers*). By 1966, there were only 8,130, less than half of the pre-war workforce. Additionally, their average age increased from 46 to 56 years.[168] Naturally, a

[167] Departmental Archives of Gironde, Pine Institute Archives, file "Institut du Pin, Ve Plan". "Recherche universitaire hautement spécialisée, construction et équipement de l'Institut du Pin", 25 February 1965.

[168] "Conseils et Commentaires dou Machinayre. En marge du Congrès des gemmeurs d'Ygos", *Bois et Résineux*, 10 February 1966.

decrease in production of end products followed. In 1970, the French manufactured 3,800 tons of turpentine spirit and 14,500 tons of rosin, while the national consumption was 15,000 and 32,000 tons, respectively, meaning that France became a net importer of resinous products.[169]

The challenges faced by the resin industry became a focal point of political campaigns during the 1969 presidential election. Georges Pompidou, the eventual president, made a commitment to address these issues should he be elected to office.[170] In the following year, in 1970, one of the Institute's main chemical engineers, Roger Bentejac, wrote optimistically: "[W]e know that resinous products are going through a crisis, but one should not believe that the situation is hopeless," and suggested that plenty of interesting research was being conducted on the improvement of resin products.[171] In the same internal report, Bentejac suggested that the "new mode of gum fractionation" (the same that had been developed in the 1940s) might save the industry. Unfortunately, these hopes did not materialise. In the interwar period, France produced 20 per cent of the global supply of resin. In 1977, this number fell below 2 per cent with 11 million litres.[172] The production moved to China and to the Eastern Bloc, as the salary costs were the most important factor in determining the resin prices. In 1977, Bentejac, in an addendum to his previous report, completely changed his tone: "All those who are somehow attached to the gum production business hope, against all

---

[169] Departmental Archives of Gironde, Pine Institute Archives, file "Saint-Symphorien". R. Bentejac, "La gemme, sa production, son utilisation", 1970.

[170] Editorial, *Bois et Résineux*, June 1969.

[171] Departmental Archives of Gironde, Pine Institute Archives, file "Saint Symphorien". R. Bentejac, "La gemme, sa production, son utilisation", 1970.

[172] In 1977, the market share of crude resin was as follows: China (31%), URSS (31%), Portugal (18%), Mexico (8%), Spain (6%), USA (4%), France (2%). However, this list does not include Greece, South-East Asia, and Eastern European countries (notably Poland), which had some share of the market as well. In other publications from the period, it was often pointed out that the data from the Eastern Bloc were not reliable. See more in Departmental Archives of Gironde, Pine Institute Archives, file "Saint-Symphorien". R. Bentejac, "La gemme, sa production, son utilisation", addendum 1977.

logic, that this activity will be not only maintained but even increased."[173] But for Bentejac, the activity was definitely dead in France.

Even though the Institute was still prosperous in the late 1960s, the brutal collapse of the entire industrial sector eventually reached it as well. The government ceased the subsidies, deeming the situation hopeless, and the local companies were unable to offer any support to the Institute because they themselves were struggling to survive. Desperate, in 1969, Brus was determined to close the Institute and fire all its employees. He retired the same year at the age of 70 [Krasnodębski, 2014]. And yet, the history of the Pine Institute continued. Brus was replaced by Jacques Valade (future French minister of research), who chose to reconstruct and reorient the Pine Institute by enlarging its purview.[174] The paper industry, pharmaceutical industry, Saint-Gobain, and aerospace industry became some of the new partners of an institution that took pride in the 1970s in its research on cellulose and various natural polymers. The Institute did not completely abandon resin chemistry, but its importance dwindled over time. One way or another, the Pine Institute after 1970 became a very different institution, sharing the name but not the profile of its predecessor. For 70 years, between 1900 and 1970, the resin laboratory of Bordeaux delivered expertise in the field of resin chemistry. The disappearance of the resin industry in the region brought this fascinating chapter of the French history of science to an end.

Conclusions: Crisis Science

The economic history of the resin industry in the Landes was full of ups and downs. After a short upsurge in the 1860s during the

[173] Departmental Archives of Gironde, Pine Institute Archives, Cahier "Saint-Symphorien". R. Bentejac, "La gemme, sa production, son utilisation", addendum 1977.

[174] Departmental Archives of Gironde, Pine Institute Archives, box "Divers". Leaflet "50ème anniversaire de l'Institut du Pin", 1987, p. 5.

American Civil War, the French resin industry had plunged into crisis. Over the following decades, periods of prosperity interspersed with periods of misery. At the beginning of the twentieth century, social and economic tensions were exacerbated, and to appease this dangerous situation, the Bordeaux Faculty of Sciences, at the demand of the local authorities, intervened and established a new Laboratory of Chemistry Applied to the Resin Industry with Professor Vèzes appointed as its director. The laboratory slowly took root in the local economic environment in the early nineteenth century, but when resin prices started rising after 1907, along with the forest workers' salaries, the laboratory that was supposed to fight the crisis was rendered obsolete. Its budget and scientific production shrank in the second decade of the twentieth century.

In 1921, a sharp drop in prices sobered up the local community. Local producers became more receptive to the voices of Vèzes and Dupont, who advocated in favour of the new Pine Institute. Alas, with the return of high resin prices, the Institute appeared to slowly once again lose the support of the local industry, which was less interested in funding science in the times of economic prosperity. After all, why invent new products if resin sold so well on its traditional markets? The Institute's resin laboratory struggled to find funding until 1927, when a new crisis, this time a permanent one, struck. From 1927 onward, resin production slowly but continuously diminished, and the position of the resin industry in Aquitaine deteriorated. This decline became particularly steep in the 1960s, and the industry itself disappeared altogether in France in the 1970s. Over that same period, the Pine Institute flourished, with budgets rocketing sky-high thanks to the subventions meant to reinvigorate and save the resin trade. Desperate companies and forest workers lobbied the government, which provided the Pine Institute with gargantuan sums in the hope of saving the business by inventing new applications for resinous products. In the end, the industry fell, but thanks to its scientific position, up-to-date equipment, and a young and dynamic team, the Institute survived and remained one of the key industrial players in Aquitaine until the 2000s, although with less emphasis on pine products.

What is remarkable is that the inverse correlation between the Institute's funding and the industry's prosperity spans different periods. It is unrelated to broader national science policies before World War I, in the interwar period, under the French Fourth Republic and under President Charles de Gaulle. This long-term funding stability can be explained by the longevity of the "local production system" of the Landes, whose collective expectations made the Pine Institute possible.

Let us review the history of expectations in the Landes region. When, in the early 1900s, protesters rallied in the middle of *la forêt landaise*, the public authorities had already run out of conciliatory measures, and a peaceful dialogue seemed unlikely. However, symbolically, science offered an unanticipated solution. Investing in a new laboratory was a tangible, uncontroversial political action. Expectations surrounding techno-scientific progress allowed for the construction of a larger political platform made of local communities. Twenty-five years later, the story repeated itself when a crisis struck Aquitaine again in 1927. Otherwise, competing stakeholders were convinced that investing together in a university laboratory would bring beneficial results to all, and the national government supported this narrative as well.

Nevertheless, helping the Pine Institute was a sign of the lack of other potential solutions rather than the result of any long-term planning. The same logic pushed the U.C.R., representing the entire local production system, to fund resin chemistry. This is one of the key elements that contributed to the Institute's success: from the very beginning of its existence, the U.C.R., being a meta-syndicate of forest workers, forest owners, and resin producers, was a profoundly divided institution made up of hundreds of individuals. Internal struggles between its various stakeholders over price regulation and benefit distribution pulled it apart at the seams, and any agreement regarding policies capable of preventing the aggravation of the economic situation was impossible to reach. The only common policy that this fragmented community could agree on was the necessity of the investment in science.

Sociologists of expectations point to the fact that "expectations link technical and social issues, because expectations and visions refer to images of the future, where technical and social aspects are tightly intertwined" [Borup *et al.*, 2006, p. 285]. Expectations about the Institute's supposed abilities to miraculously solve deeply buried economic difficulties and underlying social conflicts via technical solutions drove the emergence of fragile political and business alliances that first funded the resin laboratory and then the Pine Institute. However, precisely because of the incoherence of all these groups and the lack of common ground, these expectations were vague at best, and hardly any accountability was possible.

What sets this account apart from other similar narratives on expectations (*e.g.*, biotechnology [Fortun, 2001] or nanotechnology [Selin, 2007]) is that there was no moment of "dismay" or "toning down". Their vagueness made the Pine Institute scientists capable of reviving hope with their promises multiple times. The Institute overrode disappointments with new promises. It should not be forgotten that the name given by Vèzes to his institution, the Laboratory of Chemistry Applied to the Resin Industry, clearly suggested its application-focused profile. Twenty years later, Dupont further emphasised that the Pine Institute, contrary to the existing "university resin laboratory" (the full name was dropped by Dupont on purpose), would be application oriented [Dupont, 1921]. In 1960, the same rhetorical turn took place when the necessity of creating a new technical service was defended, as the Pine Institute was suddenly judged as being too "basic research-oriented" [Rusterholz, 1961]. In other words, the Institute was continuously reinventing itself and trying to show that with new funding, it would (finally!) be properly equipped to bring to fruition practical solutions.

The broad coalition of forest workers and resin still owners contributed generously to the Pine Institute over the period of 70 years (1900–1970) because they all shared the same intrinsic expectations about the power of science. Facing an economic crisis, they maintained their faith in the chemists from Bordeaux, who promised to solve the industry's difficulties through scientific research as long as they received a constant stream of funding.

Science made incredible advancements possible in the general understanding of the composition of resins — it offered radically new extraction methods and elaborated uses as sophisticated as rocket fuels, capturing the imagination of the general public. However, in this narrative, the Institute was never held accountable for its failures. When faced with criticism, the response was always the same — the Institute needed more money to confront real-world issues. If science did not deliver its promises, it was simply because it lacked funding or the industry was recalcitrant towards new technological solutions.

The aim of this chapter was not to suggest that resin chemistry contributed little to the resin industry. On the contrary, it is most likely that the Institute's scientific spillovers benefited a wide range of companies in France and beyond by advancing the applications of resins. Whether the work of the Pine Institute became useful in the long run at some point in history for some companies elsewhere in the world is not relevant for this case. The key observation is that an otherwise healthy, successful, and exemplary scientific institution benefited from the economic crisis and from the financial difficulties of a community that funded it for many years. Its history questions the viability and the universality of the innovation paradigm and formulates a whole set of new questions, *e.g.*, How do scientific institutions adapt their rhetoric to obtain funding during times of social and financial crises? How do fears, hopes, and expectations affect choices of funding agencies under exceptional political and economic conditions? How do crises perpetuate certain funding patterns? Moreover, this case study raises the critical question of whether the willingness of social and political actors to shoulder the additional financial burden of funding research is a sign of political wisdom and foresight or a symptom of desperation and helplessness in the face of economic difficulties.

All these questions stem naturally from the rich history of the Pine Institute and resin chemistry in Aquitaine. But they do not exhaust the topic. New light can be shed on this complex narrative if we contrast it with the disciplinary and institutional matrix of the largest resin products producer in the world: the United States.

Chapter 5

BETWEEN COLLABORATION AND COMPETITION: THE FRENCH AND AMERICAN RESIN CHEMISTRY THROUGHOUT THE TWENTIETH CENTURY

> "My luck, hit grows with de piney wood
> An' while pines grow my luck stays good
> Food in de kitchen and times ain't hard,
> When a man works out in God's front yard."
>
> Nicke, W. and DeMert, L. (1942). De woods of pine: a song dedicated to the Pine Institute of America, *Forest Facts*, 14.

The introductory quote comes from an American journal, *Forest Facts*, and is one of many examples of the pervasive influence that the pine industry exercised on the culture and identity of the American pine regions. In fact, this industry shared many features with that of the French forests on the side of the Atlantic Ocean. What distinguishes the French and American resin industries from those in Eastern Europe and Scandinavia is the importance of turpentine.

In the latter, pine tapping barely existed until the twentieth century due to a lack of know-how and less favourable climatic conditions. The main resinous products in Eastern Europe and Scandinavia were, for long, byproducts of pine wood carbonisation, such as pitch, tar, and similar substances. In the United States (US) and France, the practice of pine tapping and then gum resin distillation for turpentine opened a vast array of new market outlets in the early nineteenth century. From a technological and economic standpoint, the two industries developed along similar lines.

And yet, the American and French industries and their scientific infrastructures were not at all alike. The French Pine Institute was a local scientific institution funded by companies, local authorities, and labour unions. Its goal was to help in the sustainable development of the economy of the Landes, which relied on the artificial and carefully maintained forest. In contrast, the American scientific institutions were almost entirely funded by the federal government and their history was shaped in a radically different natural environment. Virgin forests in the United States spanned over hundreds of hectares and were exploited for rapid profit. As a consequence, the resin companies there were much less rooted in the regional socio-economic landscapes. The most striking example of this disconnect with the local community is the case of slavery (or later quasi-slavery through peonage contracts), which was widespread in pine plantations.

Despite environmental and social differences, the Americans admired the French forest management organisation and tried to emulate its successes on their own soil. The history of the circulation of ideas between the two countries is complex and multilayered. This chapter explores this history in chronological order. First, the origins of the American naval stores industry will be presented, followed by its transformation in the 1920s and 1930s, led by the French example. Thereafter, the evolution of its scientific institutions over the same period and the importance of the private sector in reshaping the American resin trade after 1945 will be examined. The chapter concludes by explaining the reasons behind the decline of the naval stores industry.

1. The Origins of the American Naval Stores Industry

The American naval stores industry was the largest global producer of resin derivatives through much of the nineteenth and twentieth centuries. This industry was also one of the most distinctive features of the economic landscape of the American South. It shaped the identity of the region's inhabitants and played a strategic role in its development. Unlike the Scandinavian and French industries, which developed organically since the Middle Ages, the American industry was much younger and only became technologically "modern" at the beginning of the twentieth century.

Early Days (Before 1900)

The history of the resinous products industry on the American continent dates back to the seventeenth century. In 1606, the first French colonists produced pine tars in the forests of Nova Scotia, but the importation of resinous products from North America gave no economic advantage to France, and the industry did not take root [Harmon and Snedeker, 1997, p. 146]. Two years after the French attempt, a group of Polish and Dutch colonists was deployed in Virginia to manufacture tars, pitch and soap from pine wood for the British empire. Once more, the costs of the operation became too high. There were few pines in Virginia, and their quality was mediocre. The conflicts with First Nations prevented the exploitation of pines farther from the coast, and, most importantly, the transportation infrastructure on the land was non-existent [Gamble, 1921, p. 21].

While the French could produce resin domestically, the British had no pine forests on their territory and had to import the naval stores indispensable for the maintenance of their fleet from Scandinavia and Eastern Europe.[175] This left Britain vulnerable in

[175] The term naval stores stuck to all resinous products in the English-speaking world even though only a fraction of resins were actually used in shipyards since the nineteenth century.

the seventeenth century, as the Dutch navy's rapid expansion led to increases in tar prices. Additionally, Denmark raised taxes on goods crossing the Kattegat sea areas, while the Hanseatic League attempted to block trade with Russia altogether. To secure raw materials for their navy and gain independence from foreign importations, the British decided to create from scratch a full-fledged naval stores trade in their colonies. In 1704, almost a century after the failed attempts in Virginia, the British Parliament voted in the famous Act for Encouraging the Importation of Naval Stores from America. The impact was immediate. In 1701, only 1 per cent of the resinous products imported to the British Isles came from the American colonies. This number rose to 25 per cent in 1714 and up to 90 per cent in 1720 [Outland, 2004, p. 13]. Suddenly, the British empire became a net exporter of naval stores products, weakening the position of countries such as Sweden, which had been relying on tar exports for decades. Over the same period, the French also tried to relaunch their own forest projects in Louisiana but failed again due to the fact that other agricultural products, such as tobacco and indigo, generated considerably higher profit margins.

The resin industry in America developed thanks to the presence of an enormous virgin forest in the southeastern part of the continent, which was dominated by longleaf pines (*Pinus palustris*) and smaller slash pines (*Pinus elliottii*). This American pine forest amounted to around 38 million hectares, so it was 380 times bigger than the Landes Forest (100,000 ha) before the appearance of nineteenth-century plantations. By the 1760s, the centre of the industry was in North Carolina, which had "only" two million hectares of forest. The entire economy of the state relied on naval stores production. In fact, in the first half of the nineteenth century, almost 100 per cent of American resins came from North Carolina and then South Carolina [Prizer, 2009, p. 38].

What were the market outlets? Turpentine spirit became widely coveted by the American producers of rubber (as a solvent) and of "camphene" (mix of turpentine and alcohol) lamps [Outland, 2004, p. 38]. These new uses encouraged the installation of distilleries on American soil. In the early nineteenth century, the Americans exported mostly raw pine resin, but by 1844, almost half of the resin was distilled into turpentine and rosin before being exported. Up to 50 per cent of all the exported resin and resinous products were consumed by three European countries: England, Germany, and the Netherlands, both for naval uses and the nascent chemical industry [Gamble, 1921b]. North Carolina's Wilmington port was the global centre of the resin trade.

As noted above, what differentiated the French and American resin industries was the fact that forest workers in the US were, above all, slaves. In fact, the cotton industry aside, this trade was one of the largest branches of the economy that relied on slave labour [Prizer, 2009, p. 38]. The work conditions did not drastically change after the introduction of the 13th Amendment to the American Constitution. The African-American labourers worked in difficult conditions, remained underpaid, and were usually at the mercy of former slave owners who often exercised political and administrative power in the regions in question. Many forest workers lived in a state of quasi-serfdom for decades, including a number of new European immigrants from Ireland and Eastern Europe. Some resin extraction operations deep in the forest were organised in a feudal manner almost until the 1940s [Outland, 2004, p. 247]. This parenthesis on the work conditions serves as a reminder that, in spite of persisting difficulties, French forest workers had considerably more agency and could influence the development of their industry.

Another important difference between France and the US was exploitation methods. The French used Hugues' pots and developed a system that allowed a single tree to be exploited for up to 70 years — four years of tapping a single face, then one year of rest, and then the opening of another face on the opposite side of the tree. In general, the French forest was managed like a factory, with constant maintenance, replanting, and road network improvements to facilitate transportation and to prevent fires. In contrast, the forest in the US was natural and seemed inexhaustible. Forest workers were unaware of the sustainable harvesting techniques employed by their counterparts across the ocean, and the operation managers showed little interest in implementing such practices. The prevailing notion was that forests had an unlimited supply, and there seemed to be no economic incentive to preserve them. Naval stores operators rented or bought vast wooded areas and then extracted as much resin as possible. The rudimentary and excessive methods killed trees after a few years of intensive exploitation. In particular, no clay pots were used, but the workers carved a so-called box in the tree itself to gather resin, a system that was both wasteful (as the box was hard to empty) and permanently damaging to the tree. The accumulation of water in the cavity caused the wood to rot, the traces of resin caught fire easily, and the trees were much more vulnerable to diseases and insects. When the trees were dead, they were cut and sold, the lands were abandoned, and the naval stores operators moved to untapped areas, leaving behind a barren territory covered with dead pine stumps. When North Carolina's forests were exhausted in the 1870s, the so-called "tar heelers" moved to South Carolina. When nothing was left in South Carolina, they relocated their businesses to Georgia and Florida. Some ignored the scale of deforestation and believed that forests regenerated rapidly and North Carolina would remain the industry's centre, but this never happened [Gamble, 1921c, p. 34].

The devastation of the southern forests became increasingly apparent not only to American foresters but also to the resin industry stakeholders themselves. One of the most vocal proponents of the modernisation of the industry was the historian and politician Thomas Gamble. He was the editor of the *Naval Stores Review*, the leading publication in the field for the entire twentieth century. In 1921, he edited the most complete book on the state of the American resin industry, titled *Naval Stores: History, Production, Distribution and Consumption*. It is a compilation of articles written not only by Gamble himself but also by scientists, entrepreneurs and foresters. Commenting on the destruction of the Carolinas' forests, he wrote:

> The exhaustion of the South Carolina pine forests, so far as heavy supplies of naval stores were concerned, was astoundingly rapid. Such a thing as conservation was undreamed of. The vast forests of Georgia and Florida and Alabama were too inviting to promote the

> thought of care in the use of what remained of the Carolina pine forests that had evoked the admiration of the early discoverers and explorers. No section of the primeval longleaf pine forests was more quickly or more effectively obliterated than the one through which the "Tar Heelers" pressed on their way from North Carolina to Georgia. A very few years and they had cut their last boxes, hacked their last trees, gathered their last crops of crude gum, and, like an army of locusts leaving a Kansas wheat farm, moved on to fields new and pastures green [Gamble, 1921d].

While the destruction of the forest was sometimes presented as a positive phenomenon as it created potentially arable area, many were aware of the fact that if the resin industry was to survive, a profound transformation was necessary. The inspiration for those trying to save the American naval stores industry came from its largest competitor: the French Landes Forest.

French Roots of the American Scientific Forest Management

The first scientific studies of the American pine forest were conducted by the French botanist François André Michaux. His father, André Michaux, was a well-known French traveller sent by Louis XVI to the US in 1785 to search for new and interesting species for the royal *Jardin des Plantes*. His son accompanied him throughout the voyage. They returned to France after the Revolution, but after his father's death in 1802, young Michaux travelled back to America to study its forests. In 1810–1813, he published a voluminous treaty titled *Histoire des arbres forestiers de l'Amérique septentrionale, considérés principalement sous les rapports de leur usage dans les arts et de leur introduction dans le commerce* (History of the forest trees of North America, considered principally in relation to their use in the arts and their introduction into commerce). It was, most likely, the first full-fledged scientific publication on the naval-stores-producing trees, but it paid no attention to the technical details of resin extraction and treatment [Gamble, 1921e, p. 9].

The French techniques came into spotlight in the US more than a hundred years later, thanks to the works of Charles H. Herty, an American chemist considered today to be the founder of resin chemistry and the modern paper industry in the US [Reed, 1995] (Figure 47). In the 1890s, during his stay in Germany, he asked his chemistry professor, Otto Nikolaus Witt, what he thought of the American naval stores industry. Witt answered him abruptly: "You have no industry; you have a butchery. I speak from personal knowledge, for I have been in Florida. You are wasting your natural resources and get nothing like an adequate return from them." [Veitch, 1921, p. 135] Witt strongly criticised the destructive practices of the American naval stores industry and believed that the American forests could be utilised in a far more beneficial manner.

Herty personally visited the forests in Georgia and Florida and agreed with Witt's assessment. He realised that far-reaching changes

Figure 47: Charles Herty (1867–1938). Chemist, businessman and president of the American Chemical Society.

Source: Library of Congress, Prints and Photographs Division Washington, D.C. 20540 USA. http://hdl.loc.gov/loc.pnp/ggbain.3246

were necessary and worked together with the United States Department of Agriculture to implement improvements. Herty's involvement in the reform of the American resin industry was thoroughly described in the publications from the period, including the fact that he was inspired by the French methods [Veitch, 1921]. What was missing was the French perspective on the matter. In fact, in 1903, Herty visited Bordeaux. Maurice Vèzes explained that his host came to France to study "the method used by our *résiniers* for harvesting resin; a method that seems particularly perfected when compared with the barbaric technique still used in the southern states." [Vèzes, 1903] Herty, according to Vèzes, went back to the US with a "firm intention to organise, with the support of his government, a laboratory and teaching program analogous" to the ones in Bordeaux.

After Herty returned home, inspired by the French technical solutions, he praised the Hugues' pots [Herty, 1903]. However, he also noted that their direct introduction to the American forest would have been a failure, since French forest workers had skills and tools that were not available in the southern states. Their implementation in the US would have been expensive and time-consuming. He underlined notably that the shape of faces in the American forests was different (due to the use of different tools) and that the American receptacle needed to be properly adapted.

Herty called his invention a cup and gutter system. He was not the first one to attempt to introduce a new harvesting method in the US, but it was his invention that gained traction.[176] Herty's device consisted of two little gutters that let resin flow from the entire exposed part of the wood (larger than in France) into a cup (Figure 48).

The introduction of the new system met some resistance, as American forest workers complained about "flowerpots". However, in the season of 1908–1909, 14 per cent of the resin in the US was harvested using Herty's method [Veitch, 1921]. In 1909–1910, it reached 25 per cent, and by 1919, it reached 80 per cent [Outland, 2004, p. 222].

[176] Already in 1869, a certain Pudigon tried to introduce resin pots in the US. In 1895, another attempt was made by a German J.C. Schuler, with no success. See [Dunwody, 1921].

Figure 48: Left: Cup and gutter system. Source: United States Library of Congress, Prints and Photographs division under the digital ID fsa.8b29660. Right: The French system. Source: "Pot de Hugues French Harvesting System." Larrousiney-private work (CC BY-SA 3.0)

Even though the cup and gutter system was a revolution, its overall importance must be nuanced. Of course, Herty's technique increased the yield and made the trees more resistant to fire and diseases. However, it did not affect the general forest exploitation policies. The trees remained tapped and exploited to death, even though it now took more time. The French advantage did not stem from a simple resin-collecting device but from an entire system taking into account forest preservation. Historian Robert Outland explained the sentiment reigning in the region in the following way:

> [The French] accomplishments amazed American producers and forestry researchers. Where France had created a highly successful naval stores industry from a once-barren sand region, the American South had accomplished the opposite, transforming a healthy pine forest into a near-worthless wasteland. Moreover, the southern United States possessed more environmental advantages — better soil, longer growing season, and more plentiful rain — than the Landes region, but was still outpaced. [Outland, 2004, p. 158]

The American scientists in the first decades of the nineteenth century deplored that there were, *a priori*, no reasons for such a disadvantage. Inman Eldredge, a forestry engineer from Florida, wrote in 1921: "The naval stores industry in France, while much younger than ours, has reached a state of refinement far beyond anything that we have dreamed of in this country, because it has developed along scientific lines." [Eldredge, 1921, p. 168]

Eldredge praised the French forest workers for being trained in the "science" of tree conservation. Not only that, he emphasised that the quality of life of the resin worker played a significant role as well: "There is a first-class school nearby for his children, a church for his woman folk, and a comfortable café where he may spend an occasional hour with his fellow *résiniers* over a glass of the good red wine of the country, damning the *Boche* [a disparaging term to describe the Germans] and criticizing the government." [Eldredge, 1921, p. 172] This idyllic description may seem to contrast with the bleaker visions of the forest workers' struggles mentioned in previous chapters. It must be remembered, however, that Eldredge visited the region during the period of its greatest prosperity (after World War I) and compared the French forest workers there with impoverished African-American workers on peonage contracts.

In 1920, the US Senate investigated the problem pertaining to the southern forests. An alarming report warned that if the American forestry policy did not change, naval stores would disappear within ten years in the US. The report also pointed out that the French would be the ones to benefit the most if this happened. The entire report, translated by Marcelle Barraud, was published in the pages of *Bulletin de l'Institut du Pin* in 1924, generating some enthusiasm among French producers. According to the report, the American naval stores business was "dying" because the primitive extraction methods prioritised short-term gains over long-term sustainability. The report also explicitly acknowledged the superiority of the French approach and mentioned that in Florida, foresters tried to introduce the French methods [Barraud, 1924].

In retrospect, the analysis of the US Senate was overly pessimistic and did not account for the rapid technological improvements that were to follow. Still, in 1926, Georges Dupont remained convinced that the American forests were entirely devastated, a situation that was meant to create an opportunity for the French industry. He stated that while some other countries tried to harvest and sell resin, "the regions where the pines are exploitable are relatively small" and that "for a long time, there is no reason to be afraid of overproduction" [Dupont, 1926b, p. 2]. The following year, in 1927, the Landes fell into a profound crisis due to, among others, the overproduction of resinous products in America. This illustrates that not only was Dupont wrong, he had also underestimated the rapid industrial and scientific development in the American naval stores industry.

2. American Resin Chemistry Research Infrastructure Before World War II

In France, the institutional history of resin research is relatively easy to delineate. The resin laboratory at the University of Bordeaux was at the heart of the discipline for more than half a century. In contrast, in the US, the situation was much more complicated, with multiple, sometimes competing, laboratories and research units established in diverse political and institutional contexts. Unlike in France, where public and private sector interests were brought together in the Pine Institute, the American pine resin industry was more reluctant to fund chemical studies directly, creating a very different power dynamic.

Birth of the American Resin Chemistry (1900–1932)

In 1862, the Bureau of Chemistry was established in the US Department of Agriculture (USDA) in Washington. Its initial mission was to analyse soils, guanos, minerals, and wine.[177] Turpentine

[177] US National Archives (College Park), Records of the Bureau of Agricultural and Industrial Chemistry. H. T. Finnernan, "Preliminary Inventories Introduction".

entered its purview for the first time in 1893. In 1899, this line of studies was developed by the Bureau's laboratory of dendrochemistry. Then, in 1902, turpentine-related questions entered the perimeter of the Bureau's Leather and Paper Laboratory.[178] This laboratory transformed into the Industrial Farm Products Division in 1927.

Since the inception of pine-resin-related studies in the US governmental laboratories, there has been an administrative difficulty that the French Pine Institute has always managed to avoid, thanks to its interdisciplinary profile. In the USDA, there were two structures that considered themselves competent to study naval stores: the Bureau of Chemistry and the US Forest Service. The latter had existed under different names at least since 1881 and worked on the naval stores production since 1905.[179] As a result, research on this subject was conducted concurrently in the laboratories of these two services, focusing on various aspects such as destructive wood distillation, the use and analysis of wood turpentine in paints and varnishes, pine oil, resin harvesting, etc. There was a conflict between the two, and in an internal memo from 1913, the director of the Bureau of Chemistry accused his colleagues of "stealing" his research subjects:

> The Bureau of Chemistry has priority in the study of these problems and just as it is not considered honest for one individual to take from another his personal property, so I don't deem it proper for one Bureau of this Department to willingly and knowingly undertake to duplicate the work that the chief officers of that

[178] US National Archives (College Park), Records of the Bureau of Agricultural and Industrial Chemistry, entry 69, ARC ID# 3887383, Subject Files Relating to Naval Stores Research, 1903–1935, file "Duplicaton of work and cooperation", 00530: 052/025/00, box 6. Memorandum on the Duplication of work between the Bureau of Chemistry and the Bureau of Forestry, 18 August 1913.
[179] Georgia Historical Society (Savannah), Naval Stores Collection, M 2022, Series 1–1, file 3. "History of the U.S. Department of Agriculture".

> Bureau know the other Bureau is properly engaged in ... Even savages recognize the right of discovery or of priority.[180]

The work was theoretically divided between the forest service and the Bureau of Chemistry based on the degree of resin transformation, with the former working on wood and harvest and the latter on resin transformation.[181] In practice, encroachments continued to be regular.

Regarding the study of the properties of gums, they were undertaken for the first time by the Bureau of Chemistry in 1907. It was, however, in 1910 when the topic became the heart of its activities due to the proliferation of fraudulent practices. In fact, it was discovered over the period that around 20 per cent of all turpentine manufactured in the US was diluted with oil-based substances and that the colour of rosin was often identified incorrectly to artificially inflate the price.[182] In other words, in the 1910s, the Leather and Paper lab mainly focused on the standardisation of resinous products and the methods of detecting adulterants, mirroring the activities of the resin laboratory in Bordeaux.

In 1923, the US Congress passed the so-called Naval Stores Act that regulated all matters concerning the quality of resinous products on American territory, based on the characteristics established by the

---

[180] US National Archives (College Park), Records of the Bureau of Agricultural and Industrial Chemistry, entry 69, ARC ID# 3887383, Subject Files Relating to Naval Stores Research, 1903–1935, file "Duplication of work and cooperation", 00530: 052/025/00, box 6. Memorandum on the Duplication of work between the Bureau of Chemistry and the Bureau of Forestry, 18 August 1913.

[181] US National Archives (College Park), Records of the Bureau of Agricultural and Industrial Chemistry, entry 69, ARC ID# 3887383, Subject Files Relating to Naval Stores Research, 1903–1935, file "Duplication of work and cooperation", 00530: 052/025/00, box 6. H. Knight, "Naval Stores Investigations: Composition, Properties and Uses of Naval Stores".

[182] US National Archives (College Park), Records of the Bureau of Agricultural and Industrial Chemistry, entry 69, ARC ID# 3887383, Subject Files Relating to Naval Stores Research, 1903–1935, file "Duplication of work and cooperation", 00530: 052/025/00, box 6. Statement of projects of leather & Paper Laboratory, 1910.

Bureau of Chemistry, which was also responsible for the technical control of samples.[183] The number of samples, however, grew rapidly, exceeding the USDA laboratories' capacity. In 1928, the newly created Food, Drug and Insecticide Administration (FDIA), the FDA's predecessor, assumed responsibility for sample controls with its own section on naval stores, while the Bureau of Chemistry (that became meanwhile the Bureau of Chemistry and Soils) concentrated exclusively on research. The FDIA's laboratory of naval stores comprised four chemists, one administration employee, and special agents (a naval stores classifier and a naval stores inspector) located in Cincinnati (Ohio) and Savannah (Georgia), respectively [Weber, 1928, p. 32].

This dissociation of functions, research and control never took place in France, where the Pine Institute was responsible for both. At the same time, the difference between the two countries should not be exaggerated. In spite of the formal separation, the two naval stores laboratories (of the FDIA and the USDA Bureau of Chemistry) were located in the same building in Washington, D.C. and headed by the same person, Fletcher Pearre Veitch, a graduate of Maryland State College and George Washington University, who had been working for the Bureau since 1904.[184]

The research work conducted by the Bureau's naval stores lab aroused the interest of the industry as early as in 1915. Thanks to lobbying, the US Congress attributed a subvention of $5,000 specifically to the research on resinous products. This subvention was renewed annually. It reached $10,000 in 1917, $15,000 in 1925, and $20,000 in 1927.[185] The funding was mostly spent on salaries.

[183] US National Archives (College Park), Records of the Bureau of Agricultural and Industrial Chemistry. H. T. Finnernan, "Preliminary Inventories Introduction".

[184] S National Archives (College Park), Records of the Bureau of Agricultural and Industrial Chemistry, entry 69, ARC ID# 3887383, Subject Files Relating to Naval Stores Research, 1903–1935, 00530: 052/024/00, box 4, file "Memoranda July 1, 1930 — June 1933". Letter from W. W. Skinner to F. P. Veitch, 10 November 1932.

[185] US National Archives (Atlanta), RG310, Agricultural Research Service, Publication Records Set, 1907–1939, Record Copy of Station Publications 101–154. G. P. Shingler, "Why a Naval Stores Experiment Station Was Established at Olustee, Florida", *The Southern Lumber Journal*, December 1932.

The laboratory had, around 1921, four or five permanent employees.[186] It should be remembered that the Pine Institute also had around five permanent researchers (administrative staff aside) over this period. At first glance, the difference between the French and American institutions in terms of the number of employees may seem marginal, but an important distinguishing factor was that, in 1927, the French institution hosted no less than thirteen students in addition to its normal researchers. The American lab was a small federal laboratory within a larger structure and had little connection to academia.

Overall, due to the differences in size, funding, and position in the scientific landscape, the Americans lagged behind their French counterparts and held their work in high esteem. When in 1924, a group of Savannah producers travelled to Bordeaux to discover the French industrial organisation, Veitch sent one of his chemists as well so that he could update the state of knowledge with the latest results coming from the French laboratories.[187]

This, however, does not mean that the naval stores laboratory of the USDA had no important scientific achievements of its own. In a report from 1921, we learn that the Bureau contributed to the development of new techniques for refining wood turpentine and new distillation methods, as well as to the standardisation of rosin grades.[188] Contrary to the Pine Institute, the Bureau of Chemistry was also involved in purely economic studies and tried, for example, to understand an abnormal rise in the turpentine markets in 1918

---

[186] The entire Leather and Paper Laboratory had ten employees. See US National Archives (College Park), Records of the Bureau of Agricultural and Industrial Chemistry, entry 69, ARC ID# 3887383, Subject Files Relating to Naval Stores Research, 1903–1935, 00530: 052/024/00, box 8. "Results prior to 1921".

[187] US National Archives (College Park), Records of the Bureau of Agricultural and Industrial Chemistry, entry 69, ARC ID# 3887383, Subject Files Relating to Naval Stores Research, 1903–1935, 00530: 052/024/00, file "Memoranda — to July 1927". Letter from F. P. Veitch to W. W. Skinner, 5 April 1924.

[188] US National Archives (College Park), Records of the Bureau of Agricultural and Industrial Chemistry, entry 69, ARC ID# 3887383, Subject Files Relating to Naval Stores Research, 1903–1935, 00530: 052/024/00, box 8. "Results prior to 1921".

(most likely due to speculations).[189] In 1928, Veitch presented the profile of the laboratory in the following way:

> The research naturally divides itself into three broad lines. First, full scale still operation for the purpose of developing the best methods of handling and clarifying the crude gum, preventing losses in turpentine and rosin.... Second, the study of conditions at individual stills and the adaptation of better methods and processes to fit the needs and requirements of each still. And, third, research at headquarters in Washington on the composition of turpentine and rosin, and the study of difficulties arising in their utilization [Veitch, 1928, p. 5].

The laboratory's most ambitious project was undoubtedly the modernisation of the industry through systematic and regular guidance.[190] The so-called Naval Stores Demonstration Work launched in 1916. The project consisted in sending experienced engineers and technicians to hundreds of workshops on the American territory. The team was tasked with teaching workshop owners how to regulate the still's temperature correctly to obtain better colour grades, how to avoid losses due to turpentine evaporation, and how to correctly permeate resin barrels.[191] USDA agents travelled every year to different regions to give counsel and suggest improvements. In total, no less than 200 distilleries benefited, free of charge, from this programme.

---

[189] US National Archives (College Park), Records of the Bureau of Agricultural and Industrial Chemistry, entry 69, ARC ID# 3887383 Subject Files Relating to Naval Stores Research, 1903–1935, 00530: 052/024/00, file "Memoranda — to July 1927". Letter from F. P. Veitch to C. L. Alsberg, 5 July 1918.

[190] US National Archives (College Park), Records of the Bureau of Agricultural and Industrial Chemistry, entry 69, ARC ID# 3887383, Subject Files Relating to Naval Stores Research, 1903–1935, 00530: 052/024/00, box 8. "Results prior to 1921".

[191] US National Archives (College Park), Records of the Bureau of Agricultural and Industrial Chemistry, Entry 73, ARC ID# 3887731, Naval Stores Research Division: Records Relating to Naval Stores Patent Applications, 1934–1941, 00530: 052/028/99, box 6. "Naval Stores Demonstration Work", 6 March 1916.

Another important aspect of the researchers' work was, of course, publications. Between 1907 and 1933, the USDA published 103 articles on naval stores, coming from the Bureau of Chemistry and the Forest Service. It is, however, important to underline that only a handful of them concerned resin chemistry as such, but much more frequently, they were informative brochures on certain practices and technologies of the industry (*e.g.*, how to charge and discharge a turpentine still, how to operate a still with and without a thermometer, *etc.*).[192] The American chemists were, above all, interested in downstream practical problems and less on chemical research, strictly speaking. Nevertheless, they kept a close eye on what was happening abroad; for example, when solicited for book recommendations on naval stores chemistry, the USDA Bureau of Chemistry would recommend their French counterpart's monumental 1924 compendium *Résines et Térébenthines* (by Vèzes and Dupont) as well as the 1928 German reference in the field, *Die Natürlichen Harze* (by Hans Wolff). No major publication on these topics in English appeared until the 1930s. Moreover, in a 1932 internal report, the Bureau of Chemistry applauded the work of the Pine Institute, stating that whereas the foundations of resin chemistry had been laid by researchers from the German-speaking world, such as Ružička, Levy, Aschan, and Fahrion, it was the French scholars who systematised the knowledge in the field:

> [T]he bulk of general systematic research on the subject of turpentine resin in all its phases has come and is coming from the French Pine Institute Laboratory work by Vèzes, Dupont, with Rouin, Dubourg, Desalbres, Michaud, Uzac and others, and to the Pine Institute chemists has fallen the task of correlating, coordinating and summarizing the accumulated information, which, if it were

[192] US National Archives (College Park), Records of the Bureau of Agricultural and Industrial Chemistry, entry 69, ARC ID# 3887383, Subject Files Relating to Naval Stores Research, 1903–1935, 00530: 052/024/00, box 4, file "Memoranda 1927 — June 30, 1935". "Publications from 1907 to 1933".

> not for their careful and painstaking literature treatment, would be a mass of confusion.[193]

Why did the Americans fail to undertake more theoretical and programmatic studies on pine resin that would have been able to compete with the ones stemming from the German and French-speaking scientific communities? Or, to put it differently, why did resin chemistry struggle to turn into a proper discipline in the US in the 1910s and 1920s? In principle, the American position might have been more favourable to fundamental research. Unlike Dupont and Brus, who had to constantly convince a broad coalition of stakeholders to fund their work, the laboratory of naval stores was a federal institution safely integrated into the USDA's administration that guaranteed stable budgets. And yet, Veitch's laboratory was, in practice, under the constant scrutiny of his superiors, unlike the Pine Institute, whose directors unilaterally decided on expenses and on the evolution of research themes.

In 1932, the House Committee on Appropriations formulated a series of questions regarding the laboratory's work, including: "2. What is its economic value? 3. Why doesn't the industry do it? 4. Does it conflict with private enterprise? [...] 8. Why shouldn't this particular work be done by universities or colleges where there are laboratories and facilities?"[194] Veitch had to provide thorough explanations. For example, as for the place of the American resin industry in the research, Veitch explained: "[T]he naval stores industry is carried on largely in small units, by individual farmers and small firms, and in widely scattered localities. Producers are not thoroughly organized and even if they were, they would not be able

[193] US National Archives (College Park), Records of the Bureau of Agricultural and Industrial Chemistry, entry 69, ARC ID# 3887383, Subject Files Relating to Naval Stores Research, 1903–1935, 00530: 052/024/00, box 4, file "Memoranda 1927 — June 30, 1935". Internal report, 6 February 1930.

[194] US National Archives (College Park), Records of the Bureau of Agricultural and Industrial Chemistry, entry 69, ARC ID# 3887383, Subject Files Relating to Naval Stores Research, 1903–1935, 00530: 052/024/00, box 4, file "Memoranda 1927 — June 30, 1935". Internal report, no date.

to finance any extensive research work." In fact, he went on to say that the industry was losing $12 million yearly because of obsolete technologies, and he argued that the lab's work would save an equal amount. He added that the work of the Bureau of Chemistry did not conflict with private enterprise since "the results of our work are published for the benefit of all who are interested, and progressive individuals and firms may use them to their advantage". Finally, as for the role of the universities, he explained: "[T]here are no universities or colleges which have the proper facilities. The problems are not of purely scientific nature but are more of a technological and industrial character. Their solution requires not only special large-scale equipment but personnel having background of actual experience in the naval stores industry." He did admit that the Bureau might benefit from some cooperation with state universities.

The Pine Institute of Bordeaux was never subjected to this kind of scrutiny. Such questioning would have been judged by its directors, the *grand patrons* of resin chemistry, as an outrageous attack against their own competence. The accountability logic inherent in the US was foreign to the French resin chemists, who played the role of experts and policy advisors. It was up to them to evaluate the industry's performance and governmental policies, not the other way around.

That does not mean that there were no similarities in the position of Veitch and his French counterpart; for example, the language he used to communicate with the entrepreneurs was not that different from that used in France. Notably, he warned his fellow nationals that Germany and France outpaced the US regarding the development of new applications, despite the former not manufacturing turpentine at all and the latter only producing a third of the American output.[195] In France, it was the US and Germany that continuously threatened the future of the industry and against which the industry

---

[195] US National Archives (College Park), Records of the Bureau of Agricultural and Industrial Chemistry, entry 69, ARC ID #3887383, Subject Files Relating to Naval Stores Research, 1903–1935, 00530: 052/024/00, box 4, file "Memoranda 1927 — June 30, 1935". "Industrial-Farm Products Division — Naval Stores Investigation Program", 23 March 1931.

had to mobilise itself. Overall, this type of propaganda was of smaller importance in the US, as funding came first and foremost from the federal government, thus fluctuating less than in the case of the Pine Institute, but it certainly had an indirect effect on the selection of funding priorities at the USDA.

Naval Stores Industry and the Pine Institute of America

The contrast between research infrastructures in France and the US immediately raises the question of whether the similarities and differences between them were due to the differences between the respective industries. The French industry was sedentary, located south of Bordeaux, and deeply rooted in the broader economic realities of the region. This situation facilitated the emergence of labour unions and stable economic and scientific institutions. The American industry was, for many decades, in constant movement from exploring virgin territories and abandoning them shortly after. However, in the 1910s and 1920s, largely thanks to conservation policies, the naval stores business became increasingly sustainable and took stronger roots in Georgia and Florida, with the epicentre in Savannah, where the largest American resin market was located. Smaller entrepreneurs became increasingly interested in collaborating rather than competing, especially to confront their common foreign competitors. These conditions were favourable to the establishment of more stable trade organisations.

This happened for the first time in 1906 when the American Naval Stores Company was established. It grouped together a wide range of stakeholders of the resin business [Averitt, 2007, p. 11]. Its purpose was to reorganise the entire national resin trade, but its existence was short-lived, and it failed to achieve its goals. In 1909, its creators were accused of violating the American antitrust law, the Sherman Act. This was the first criminal indictment based on the regulation in its history. The US Supreme Court overturned the decision in 1913, but between 1909 and 1913, the company effectively ceased to exist, even though, from a formal point of view, no judicial instance ever qualified the company as an illegal trust.

Meanwhile, in France, the conditions were considerably more favourable for grass-roots trade-related endeavours. In the early 1900s, an informal community of entrepreneurs began to coalesce around the Bordeaux resin laboratory, eventually forming the Pine Institute Association in the 1920s. Simultaneously, French forest workers were organising themselves into increasingly influential labour unions, enabling them to exert pressure on the direction taken by the resin industry — a remarkable development not witnessed in the US. It is worth noting, however, that the working conditions of American workers also experienced improvement during World War I and the early 1920s. Some entrepreneurs even complained that increased salaries made their employees "lazy" [Carson, 1921, p. 73].

The early 1920s was a period of growing stability in the American naval stores industry, and local producers made one more attempt to coordinate their efforts in spite of the failure of the American Naval Stores Company a decade earlier. After an American delegation visited France in 1924, they decided to emulate the French institutions and establish their own Pine Institute of America in 1925. What was its mission? One of the founders, O.H.L. Wernicke, explained: "The one big job of the South is to make better use of its idle lands." [Wernicke, 1925, p. 150] For that, the Americans needed the appropriate knowledge and knew that "knowledge and ability are neither gifts nor endowments, they must be acquired and cultivated. This can be done by organized cooperation under wise directions and skillful management". For Wernicke, the new Institute would be an "agency to be owned and controlled by those to whom it is to benefit primarily ... Like the hub of a wheel, its many spokes can be made to coordinate with every fount of useful knowledge within the radius of all organized industry including the government, our universities, and private research agencies, and bring them to bear on our particular problems". In other words, according to Wernicke, the Pine Institute of America was meant to become both a producer of new technological know-how as well as a coordinator of its diffusion — "Organized knowledge has always and everywhere been the foundation of progress in all departments of human endeavour. It will not fail us in this case, in proof we need only to remember that its pine region, now the richest of France, was malarial swamp and barren sand before organized knowledge was brought to bear on the situation less than a century ago." There were jumps in Wernicke's logic when regarding the sources of the French success, but his admiration for the French system was a clear motivation to establish the new institution.

The Institute was to be guided by a consulting committee comprising prominent scientists and private sector representatives. It was to have many departments and employees, and provide education. Wernicke explained that the Institute should be located in the forest region in order to become "a great university in the woods" [Wernicke, 1925, p. 148].

Overall, the American producers were no less ambitious than their French counterparts and conceived the Institute as the future global centre of resin chemistry and all naval-stores-related research. Its establishment was noted with great interest in France: "The Americans ... finally recognised the necessity of applying scientific methods in their resin industry that remained, until today, rather empirical. They were inspired in their project, without hiding it, by our French Pine Institute," wrote *Bulletin de l'Institut du Pin* in 1925 [N. N., 1925b, p. 27]. In 1927, Dupont became an honorary member of the Pine Institute of America [N. N., 1927] alongside Pierre Maydieu, one of the largest French producers [N. N., 1927b]. It does not mean that the relations were always cordial. In 1927, when the Americans filed a few patents on rosin saps, Dupont reacted in *Bois et Résineux* by explaining that there were at least two French patents covering the same techniques and noted that "we should not get ourselves stripped by our American friends" [Dupont, 1927c].

Despite the ambitious plan laid down for the Pine Institute of America, it was never translated into reality. Directed by Wernicke until 1927, and then by chemist Carl Speh, it remained mostly a virtual institution. It had a small office in the Barnett National Bank Building in Jacksonville, Florida, but no laboratories of its own. In practice, it funded research in other institutions.

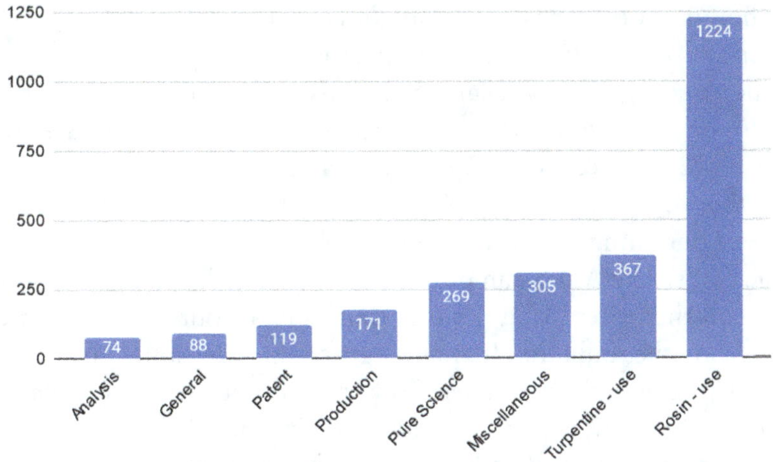

Figure 49: Number of summaries published in *Pine Institute of America Abstracts* (1927–1930).

How then did the Pine Institute of America mark the history of resin chemistry in the US? First, it funded scholarships at the famous Mellon Institute in Pittsburgh.[196] Some of the research topics were very similar to the ones studied in the French institution; for example, attempts were made to prove that turpentine is a better solvent than white spirit. In addition to funding original research, the Pine Institute of America also published a monthly publication that summarised and translated resin-chemistry-related articles published all over the world. Unlike the French *Bulletin de l'Institut du Pin*, the *Pine Institute of America Abstracts* did not publish original articles at all. The Abstracts appeared between 1927 and 1930 and had, by the end of its existence, around 300 subscribers. The summaries published on its pages were divided into eight categories: Pure Science, Analysis, Production, Rosin — Use, Turpentine — Use, Miscellaneous — Use, General, and Patent. "General" corresponded to non-scientific questions, and "Production" to distillation [Burnett and Odell, 1927, p. 5]. Figure 49 presents the respective weights of these categories.

[196]Departmental Archives of Gironde, Pine Institute Archives, box "Pine Institute". Proceedings of Seventh International Conference on Naval Stores in Florida, 1930, p. 63.

The first observation is the unilateral domination of rosins. Turpentine, occupying the second position, corresponded in terms of volume to only 25 per cent of the rosin category. In the French *Bulletin de l'Institut du Pin*, in the same period, different parts were much more balanced (see Figure 12). The second observation concerns the very small share of "Pure Science", in which appeared articles containing detailed chemical descriptions of terpenes and resinic acids. The American publication discussed, above all, applications. Another interesting point is that only 40 out of 2,617 articles came from the Bulletin. It may suggest that, in spite of symbolic gestures, the actual knowledge flow from France to the US was less important. However, it may be also argued that the Abstracts readers were mostly men of industry, and that they were less interested in more scientifically oriented publications that were, for example, read by the USDA chemists. There is no clear explanation as to why "Pure Science" was so limited in the Abstracts, especially since the industry-oriented *Naval Stores Review* followed the French institute's contributions closely.[197]

In contrast, the Abstracts was highly regarded in France. In the Bulletin, in its section "Documentation" (thus not original articles), almost half of all the summaries published between 1928 and 1930 came from the Abstracts. In other words, the Americans collected articles in the field, often from the German-speaking reviews, prepared their summaries, and then they were republished again in Europe in the Bulletin. In 1930, the Abstracts was discontinued, but in the same year, a major collection of articles selected by researchers funded by the Institute's scholarships was published as *Pine Institute Technical Bulletin No. 1: Rosin in Synthetic Resins* [Burnett, 1930]. It served as an introduction to the topic for generations of American chemists to come, and it remains the only book ever published by the Institute. No follow-up publication was ever published.

The Pine Institute of America struggled with funding in 1930, and its creators expressed disappointment that the number of new affiliates

[197] Departmental Archives of Gironde, Pine Institute Archives, file "Correspondance 1939 G.B., N". Letter from Naval Stores Review to the Pine Institute, 20 March 1939.

was not advancing as rapidly as they had expected in 1925.[198] Despite these challenges, the Institute still had a significant impact on the region. It was a member of the Southern Forest Research Council and the Industrial Committee of Florida State Chamber of Commerce. It took official stances on issues such as taxes and tariffs. The Institute was also involved in collecting data and market information related to the consumption of resinous products, and its members served on the naval stores advisory committee attached to the USDA.

In 1930, the Institute organised the Seventh International Conference on Naval Stores in Jacksonville.[199] This major event brought together key stakeholders in the field to discuss a broad range of issues, some of which are worthy of mention. Lucien Memminger, the US Consul General in Bordeaux, made a speech praising the qualities of the French forest that he felt should be emulated in the American South. He explained his role as an information broker, gathering details about the prices, production, and exports of French naval stores. Memminger had been responsible also for hosting the American delegation in 1924 in Bordeaux and explained that a new visit was being planned in 1930 but was delayed for a time because of the economic difficulties.[200] Memminger's speech also gave a clear indication that the American public was familiar with both Dupont and the Pine Institute of Bordeaux, as none of them needed an introduction. He noted, with certain pride, that the *Pine Institute of America Abstracts* was also known in France. In fact, France had its own delegates at the conference — even Dupont had wanted to attend but had to cancel his visit due to unforeseen circumstances.[201]

---

198 Departmental Archives of Gironde, Pine Institute Archives, box "Pine Institute". Proceedings of Seventh International Conference on Naval Stores in Florida, 1930, p. 259.

199 Departmental Archives of Gironde, Pine Institute Archives, box "Pine Institute". Proceedings of Seventh International Conference on Naval Stores in Florida, 1930, p. 267.

200 Departmental Archives of Gironde, Pine Institute Archives, box "Pine Institute". Proceedings of Seventh International Conference on Naval Stores in Florida, 1930, p. 17.

201 André, after his return to France, related briefly in *Bois et Résineux* all information he had acquired on the resin market in the US. See [André, 1930].

While the conference explored many issues, the future of the Pine Institute of America was also briefly touched upon. One forester suggested that the main mission of the Institute should be standardisation and knowledge diffusion, and that it should be a "trade association", not a scientific institution. He emphasised the need to prioritise "commercial research" aimed at exploring new market opportunities for existing products. It is worth noting that the opposite logic was favoured in France: market outlets were to be found through laboratory research, leading to completely new products.[202]

This practicality-oriented approach was supported by some attendees of the conference, but many chemists argued the opposite and insisted on the importance of fundamental studies. For example, Herty pointed out that chemical research was insufficient, disorganised and underfunded in the American South.[203] He cited the case of a chemist from Georgia who wanted to work on a non-saponifiable part of rosin but lacked equipment. His work was made possible by a generous donation from two widows of late resin entrepreneurs. For Herty, the reliance on private donations illustrated the weakness of the American research system. However, he also noted that the naval stores question was beginning to attract attention from academic institutions, such as the universities of Florida and North Carolina, which could lead to promising developments.

Contributions of the Pine Institute of America aside, the essential part of scientific research on resin chemistry was done by the federal administration, whose representatives also presented their contributions during the congress. It was Henry Knight, president of the American Institute of Chemists and director of the Bureau of Chemistry and Soils, who introduced the participants to the philosophy of resin chemistry research. He explained that resin-related studies were time-consuming as they had to be polyvalent and concerned different stages

[202] Departmental Archives of Gironde, Pine Institute Archives, box "Pine Institute". Proceedings of Seventh International Conference on Naval Stores in Florida, 1930, p. 60.

[203] Departmental Archives of Gironde, Pine Institute Archives, box "Pine Institute". Proceedings of Seventh International Conference on Naval Stores in Florida, 1930, p. 180.

of resin treatment. In terms of immediate results, he enumerated rosin esters developed in the Bureau of Chemistry and research on rosin crystallisation that, according to Knight, would save the industry millions of dollars.

Knight underlined, however, that if the producers wanted to see real progress, his service would need semi-industrial equipment and space. At first, the Bureau of Chemistry wanted to work with the existing companies, but the Bureau's projects often interrupted their normal operations. That was why the Bureau needed an independent research station entirely devoted to practical resin-oriented research. Knight wanted to reunite, in a single institution, basic chemical research (*i.e.*, "technical research") and forestry research. His idea started to take shape in 1931.

The Olustee Research Station

In the 1920s and early 1930s, more and more key actors of the naval stores world, including the Pine Institute of America, voiced their concerns that the Washington laboratory was too far from the practical challenges of everyday work in Florida and Georgia. In 1923, a new position — naval stores technologist — was created in Savannah to form a "permanent link" between Washington and the South, but it was far from sufficient.[204] However, the lobbying from the Pine Institute of America members in the US Congress bore fruit. A special supplement of $40,000 was allocated to the Department of Agriculture to create a proper naval stores experimental station. An advisory committee was set up to oversee the project.[205] After many consultations, the committee chose Olustee, Florida, for the location for the

[204] US National Archives (Atlanta), RG310, Agricultural Research Service, Publication Records Set, 1907–1939, Record Copy of Station Publications 101–154. G. P. Shingler, "Why a Naval Stores Experiment Station Was Established at Olustee, Florida", *The Southern Lumber Journal*, December 1932.

[205] US National Archives (College Park), Records of the Bureau of Agricultural and Industrial Chemistry, entry 73, ARC ID# 3887731, Records Relating to Applications for Patents Pertaining to Naval Stores, 1934–1941, 00530: 052/028/99, box 5. G. P. Shingler, E. L. Patton, "The Naval Stores Station, Olustee, Florida".

new station, near an experimental forest of the Forest Service. This decision was seen as an attempt to strengthen the collaboration between foresters and chemists. The new location was approved on 21 May 1931, funding was allocated a week later, and on 16 August 1932, the Olustee Experimental Station started operations (the same year, on the other side of the Ocean, the forest laboratory of Pierroton was acquired by the Pine Institute [N. N., 1932d]). Veitch rejoiced:

> The need for a naval stores station has been felt by the naval stores people for years, and it is the producers, individually and collectively, through their organization, the Pine Institute of America, that have made possible the opening of this station, the first of its kind to be established in the United States. ... Had it been opened 10 years sooner the naval stores industry would have had an easier time in recent years.[206]

The area of the station was 40,000 m^2. At first, it had but a few administrative buildings and a single still, but by 1935, it had a fully-fledged modern chemical laboratory.[207] It was a federal institution and thus financially independent from the local industry, unlike the Pine Institute. It relied, however, on free resin attributions (for its experiments) from local companies.

The Station was a matter of pride for the Bureau of Chemistry. "The naval stores station is not the usual turpentine still. It is much more. The best, most costly turpentine still in the country has neither the buildings, equipment, nor staff to carry on the scientific research work that will be undertaken here", read a brochure from 1934 [Veitch, 1934]. Overall, the budgets of the naval stores section of the

---

[206] US National Archives (Atlanta), RG310, Agricultural Research Service, Publication Records Set, 1907–1939, Record Copy of Station Publications 155–200. F. P. Veitch. "The Naval Stores Station of the Bureau of Chemistry and Soils", USDA Miscellaneous Publications, n. 206.

[207] US National Archives (Atlanta), RG310, Agricultural Research Service, Publication Records Set, 1907–1939, Record Copy of Station Publications 501–577. G. E. Hilber (1953), "Twenty Years of Research by the Naval Stores Station", *AT-FA Journal*, 15(4).

Bureau of Chemistry witnessed a manifold increase in budget during the 1930s compared to the 1920s. The budget of the entire section (the Olustee Station and the laboratory in Washington) was $59,330 in the 1935–1936 budgetary year.[208] Thanks to this ample funding, the section could employ fifteen permanent staff members at two sites: two chemists, one chemical engineer, one turpentine stiller, two clerks, two cooperative agents, and seven still workers. The expenses should not have mattered since, according to Henry Knight, the work of the Olustee Station alone would permit the economies to be as high as $1.5 million a year to the industry.[209] And yet, Veitch considered the budget far from satisfying and hoped to double it. According to him, 250,000 to 300,000 Americans depended on the resin industry, so the federal government could not afford to neglect it.[210]

While the contributions of the Olustee Station are unquestionable, its creation disrupted the power structures in the American naval stores chemistry. First, its establishment led to the Pine Institute of America losing its importance over the years. In fact, there is little trace of the Institute after 1932, and it closed its doors for good by 1936.[211]

[208] The relation between the base budget and the budget exclusively for the establishment of the new station remains unclear. In 1933, "Set-Up Naval Stores Works" were evaluated to 68,000 dollars, 70 per cent of this amount was attributed to Olustee and 30 per cent to Washington. See more in US National Archives (College Park), Records of the Bureau of Agricultural and Industrial Chemistry, entry 69, ARC ID# 3887383, Subject Files Relating to Naval Stores Research, 1903–1935, 00530: 052/024/00, box 4, file "Memoranda 1927 — June 30, 1935". "Program on Fundamental Research on Naval Stores".

[209] US National Archives (College Park), Records of the Bureau of Agricultural and Industrial Chemistry, entry 72, ARC ID# 3887730, Manuscripts Relating to Naval Stores Research, 1928–1940, 00530: 052/027/00, box 3. H. Knight, "What's ahead in naval stores", 1935.

[210] US National Archives (College Park), Records of the Bureau of Agricultural and Industrial Chemistry, entry 69, ARC ID# 3887383, Subject Files Relating to Naval Stores Research, 1903–1935, 00530: 052/024/00, box 4, file "Memoranda 1927 — June 30, 1935". "Program on Fundamental Research on Naval Stores".

[211] US National Archives (College Park), Records of the Bureau of Agricultural and Industrial Chemistry, entry 32, ARC ID# 568557, General Correspondence 1935–1939, box 26, 39029–39045, file "Naval Stores — Miscellaneous". Letter from F. P. Veitch to Allan McDonald, 28 November 1936.

In 1936, a new entity was formed: the American Turpentine Farmers Association (ATFA), whose mission was to disseminate trade and economic information concerning resinous products, but it had no intention of funding any scientific research. It closely followed, however, the discoveries and improvements of the federal laboratories and presented them in its own bulletin.[212] It looks like the Pine Institute of America disappeared after fulfilling its initial ambition, even if it did so indirectly. Clearly, the Olustee Station was not the "university in the woods", but it was the best the industry could hope for. In this sense, there are some parallels between the Pine Institute Association and the Pine Institute of America. Both entities lost their *raison d'être* after the scientists they were supposed to help had managed to achieve their goals. The Olustee Station, a federally funded entity that was created thanks to the Institute's lobbying, no longer required its assistance after securing funding in the same way as the Pine Institute of Bordeaux did not need the Pine Institute Association after the convention between the Association and the Bordeaux Faculty of Science was signed in 1937.

The establishment of the Olustee laboratory also disrupted relations within the federal research infrastructure itself. Its growing significance led to the liberation of the naval stores laboratory from the tutelage of the Industrial Farm Products Division in 1935. Veitch became director of the newly created USDA Naval Stores Research Division.[213] However, this rise to power was mainly due to the successful installation of the Olustee Station, which was far from Washington and its institutions. The director of the Olustee Station, G.P. Shingler, a former naval stores technologist who was familiar

[212] Ex. US National Archives (Atlanta), RG310, Agricultural Research Service, Publication Records Set, 1907–1939, Record Copy of Station Publications 501–577. Editorial, "Research Program for Naval Stores Station", *AT-FA Journal*, April 1955; G. P. Shingler, "Contributions of Naval Stores Research Division to the Processing Of Pine Oleoresin", *AT-FA Journal*, March 1946.

[213] US National Archives (College Park), Records of the Bureau of Agricultural and Industrial Chemistry, entry 32, ARC ID# 568557, General Correspondence 1935–1939, box 5, 38466. Letter by F. P. Veitch, 19 July 1935.

Figure 50: Left: G. P. Shingler, technical director of the Olustee Naval Stores Station. Right: F. P. Veitch, Director of the Naval Stores Division in the USDA.

Source: US National Archives (Atlanta), RG310, Agricultural Research Service, Publication Records Set, 1907–1939, Record Copy of Station Publications 101–154.

with the local producers, saw himself as the true leader of the new scientific structure (Figure 50).

In reality, a certain tension between the two directors had been discernible between them from the 1920s. Shingler complained regularly about his ambiguous title ("naval stores technologist"), which he believed was poorly understood by the industry members. He wanted to be called "associate chemist", but Veitch refused every time, most likely because Shingler had no formal chemical education.[214] After 1932, the tension between the two became more and more pronounced. Shingler wanted more independence and recognition. In 1936, Veitch had to explicitly remind Shingler that the Naval Stores Research Division had only one director — Veitch — and that

[214] US National Archives (College Park), Records of the Bureau of Agricultural and Industrial Chemistry, entry 70, ARC ID# 3887384, Records of G. P. Shingler Relating to Naval Stores Demonstration Work, 1914–1936, 052/026/00, box 5. Letter from F. P. Veitch to G. P. Shingler, 23 March 1929.

the Naval Stores Station in Olustee was part of the Washington D.C. centre and not an independent research unit as Shingler imagined. Veitch refused to establish the position of "chief" of the station because, according to him, this title required "a certain level of expertise" that no one in Olustee had. He proposed to Shingler the title of "acting chief" to underline the difference, but his offer was not taken up. Finally, Shingler managed to obtain the position of "senior chemist in charge".

Whatever these internal conflicts were, the Olustee Station had tangible scientific and technological contributions. One of the most important innovations from the 1930s and 1940s in terms of promoting the station's contributions was the development of the so-called Olustee process of gum cleaning. In France, gum had been cleaned before distillation (*térébenthinage* in French) since the nineteenth century to remove debris (bark, insects, twigs, etc.). In the US, there was no single standardised or customary process of cleaning, and many workshops distilled the polluted gum directly, which led to a very low quality of rosins in comparison to the French ones. While there is no clear answer as to why such a basic technique failed to penetrate the American industry, the low value of rosins in the US due to the lack of market outlets might have contributed to the neglect of the cleaning process. In France, rosins were long used for the manufacture of soap, an outlet that required bright rosins from thoroughly cleaned gum. But other factors might also have played a role, *e.g.*, the different viscosity of American gums [Shingler, 1948].

Prior to its implementation, approximately 80 per cent of rosin produced in the US was dark (*le brai*, according to the French nomenclature); after its introduction, 80 per cent was clear and bright (*la colophane* in French).[215] Almost all the American distilleries used the Olustee technique by the end of the 1940s. Even though it was patented, the technique was freely available to all interested

[215] Georgia Historical Society (Savannah), Naval Stores Collection, M 2022, Series 9, file "Publications — articles and Newsclippings 1930–1950". E. L. Patton, "Naval Stores", June 1948.

parties, and the Olustee technicians were available to assist with the installation of the process.

Another great contribution of the Olustee Station was the development of a new type of steam distillation still. In the US, the vast majority of workshops in the 1930s used old naked flame distilleries where turpentine was mixed with water and heated directly in the recipients, a technology that had become obsolete in France in the first decade of the twentieth century. The Olustee Station inventions and knowledge transfer activities transformed this state of matter and spread the use of steam distillation in American distilleries.

Finally, the last of the Olustee Station's great innovations was continuous distillation, a topic thoroughly studied by the Pine Institute as well. While continuous distillation technology had been introduced in France in the 1910s, it did not arrive in the US until 1940, thanks to the Olustee Station. Ironically, after World War II, the Olustee model was actually implemented in some French plants due to its perceived superiority.[216]

The long list of the Olustee Station's innovations is impressive. Nevertheless, they all have one thing in common: they served the gum naval stores industry. Over the same period, however, a different branch of naval stores started rising in popularity: the so-called wood naval stores obtained by direct pine wood distillation.

3. Reclaiming the Wastes and the Discovery of Wood Naval Stores

One of the missions of the Olustee centre was to find new uses and exploit the so-called resin dross (*griche* in French), little pieces of wood soaked in resin. Long considered as waste to be filtered and gotten rid of, special distillation columns were designed to extract

[216] US National Archives (Atlanta), RG310, Agricultural Research Service, Publication Records Set, 1907–1939, Record Copy of Station Publications 578–653. Research Achievement Sheet, "Continuous steam still produces better turpentine and rosin at lower costs".

valuable substances from the dross, turning it into a full-fledged raw material. This trajectory was widespread in the naval stores industry. Rosin and dead pine stumps can serve as examples of this phenomenon, and this section explains how these two categories of previously useless wastes became valuable products. Although attempts to find new uses for the latter were initially timid and had low profit margins, they eventually revolutionised the entire naval stores industry on a global scale.

Rosin Mines

For many years, rosin was considered a valueless, wasteful byproduct of resin distillation in the US. Although it started to substitute fatty acids in the soap industry in the early nineteenth century, the lack of railways and road networks made its transportation from the forests in the south economically non-viable, unlike the much more expensive turpentine. The rosin produced in the distilleries was simply rejected into lakes, rivers, or manually dug holes in the forest. Over the years, rosins started accumulating, forming impressive deposits. However, when the transportation network expanded by the end of the nineteenth century, and the soap and varnish industry became more interested in acquiring rosin for the manufacture of cheaper consumer products, the solid distillation product found a new life and rosins turned from a waste product into a full-fledged raw material.

Interestingly, while the naval stores industry was selling its rosins on a daily basis, it was discovered that deposits of previously discarded rosin, especially those kept safely under water where they did not oxidise, had the same quality as the rosin extracted directly from pine gum. Suddenly, these rosin deposits in the abandoned wasteland of the North and South Carolinas became true rosin mines, and many jumped onto the bandwagon of the new "rosin rush", searching the country for forgotten man-made ores. The most accessible ones on the surface were exploited first. With time, however, the explorers of "pine gold" ventured farther by looking for rosin at the bottom of rivers and lakes. Some deposits had been accumulating for no less than 70 years and were counted in thousands of rosin barrels [Gamble, 1921f].

Figure 51: Rosin Mine at the bottom of Catherine Lake in North Carolina. *Source*: [Gamble, 1921f]

Yet, this new industry did not survive long, and the man-made deposits were quickly depleted. It was in 1920 and 1921 when probably the most impressive rosin extraction operation in the history of the trade was launched at the bottom of Catherine Lake in Onslow, North Carolina (Figure 51). A certain Percy L. Gardner profited from the high prices of rosins immediately after the war to fund the construction of a dam in order to dry the lake and extract its rosin deposits.

By the 1930s, these mining operations were but a curiosity in the long history of naval stores industry. However, over exactly the same period, another former waste was elevated to a key raw material: pine stumps.

Wood Naval Stores

The distillation of pine gum into turpentine and rosin constituted the most important innovation in the entire history of the naval stores industry, whose history spanned, as we have seen, thousands of years. Before the widespread adoption of distillation in the eighteenth and nineteenth centuries, resinous tars served as the primary raw material for the industry. These tars were produced through a process involving enormous kilns made of sand and dirt, where stacks of piney wood were placed. Through pyrolysis, charcoal was produced and left in the kiln, while the thick tar flowed through designated openings into containers. This method is known as "dry wood distillation" and is the most traditional form of wood distillation, which has been practised in various forms for hundreds of years.

Although this technique developed dynamically in the nineteenth century, one invention stands out. In 1872, a certain James Stanley from Wilmington perfected a technology he called "destructive pine wood distillation".[217] Stanley used an industrial kiln to gradually increase the temperature to 200°C, which liberated turpentine alongside various other rosin oils from the wood in addition to charcoal and tar. This turpentine was of much lower quality than the one manufactured by resin distillation because of pollution, but it was a genuine alternative — a new source of potentially marketable resinous products that did not require the labour-intensive practice of pine tapping and resin collection. This simple technique was studied in the Pine Institute in the 1920s as well [Dupont, 1930]. Despite having fewer market outlets than turpentine and rosin, pine tars extracted by dry wood distillation continued to be manufactured for many years, thanks to the interesting properties that petroleum derivatives did not possess [Mims, 1949]. The raw material for destructive wood distillation was most often the so-called "light wood" (*bois gras* in French, fat wood), made of forest debris (*e.g.*, twigs, branches, cones, roots) often left on the ground for many months before being collected. Light wood, though very flammable, was much richer in resin than living pine wood. Normally, resin constitutes about 5 per cent of the weight of living pine wood, but

[217] Othmer Library (Philadelphia), Hercules Inc. Series IV — Hercules History 2, file "Naval Stores". "Naval Stores History We Never Knew 'til Now", no date.

in light wood, because of its time spent on the ground, can reach up to 40 per cent or more.

The destructive wood distillation process had the potential to utilise pine stumps, as mentioned earlier, a previously unexploited raw material. The vast territories of North Carolina and South Carolina, covered by pine forests only a few decades before, had turned into barren wastelands due to the extensive naval stores operations that effectively killed entire forests, leaving behind only pine stumps. These lands were not welcoming for potential farmers as these stumps had to be extracted from the ground to make the land cultivable, which was a time-consuming and costly operation. However, these pine stumps, like light wood, were very resinous, and this led some entrepreneurs to consider their use for naval stores extraction in a more organised manner. This new resource was never exploited in France, where, thanks to the reforestation practice, no pine stump wastelands were ever created. In France, such stumps were usually removed immediately after clearing, leaving no time for resin to accumulate. If the stumps had not been removed, they would have been surrounded by new growth after a few years, with no easy way to extract them without threatening neighbouring live and productive trees. Although the Pine Insitute chemists were aware of the high yield of resin from these old stumps, attempts to extract resinous products from them were few and far between [Dupont, 1927d].

In the US, the problem of stump extraction and the valorisation of stumps became increasingly pressing as wastelands expanded alongside the need for arable areas. In 1906, Homer T. Yaryan, an engineer from Ohio, came across an advertisement of a company extracting turpentine from pine stumps that had just started its operations. The company was using destructive wood distillation for this purpose. Yaryan thought about using a different method, so with the support of some of his friends, he launched a pilot plant in Toledo (Ohio) and set up the Yaryan Process Company, which would use a new "non-destructive steam and solvent wood distillation" process.

This technique "consists in grinding the wood by means of a hog and shredder, then steaming it to remove turpentine and some pine oil. The residue is extracted with gasoline forming a solution of rosin and gasoline. By distilling off the gasoline the rosin remains behind" [Harper, 1921, p. 245]. The operation is repeated a few times, and rosin is redistilled to eliminate residual turpentine. Its two major products, wood turpentine, and wood rosin, could now compete with the ones extracted from the gum. It is worth noting that this technique, unlike the dry or destructive wood distillation, did not produce charcoal and tar.

At first, the process posed a range of technical problems. The laboratory tests had been much more promising than the industrial-level ones. Yet, with the help of an investor who decided to plough no less than $120,000 into the project, Yaryan constructed a huge wood distillation plant in Gulfport, Mississippi, in 1911. Another plant was opened in Brunswick, Georgia. The two plants, however, used mostly light wood and not pine stumps, whose extraction turned out to be prohibitively expensive. Another problem was that the quality of the products was very low. The rosins were almost black and had very few uses. Yaryan's company only survived until 1913 because Yaryan had no sales department of his own — all sales had been handled by the American Naval Stores Company, which closed that year. Without sales, his firm fell quickly into bankruptcy. Undeterred, Yaryan established a new company — Yaryan Rosin and Turpentine Company — in 1915. This time, however, Yaryan was aware that a stable influx of cheap raw material, such as pine stumps, was necessary to turn a profit.[218] Hercules Powder Company came to his aid.

Hercules Powder Company

In 1912, the famous DuPont company was investigated by the federal government for violating the Sherman Act (the antitrust law). The courts ordered a divestment in the powder production [Posedel and Lawton, 2011, p. 25]. This led to the creation of two companies: Hercules Powder Company and Atlas Powder Company. At the moment of creation, Hercules had 1,000 employees and nine plants.

[218] Othmer Library (Philadelphia), Hercules Inc. Series III — Public Relations-1, box 5, Naval Stores History 5/4. V. R. Crosswell, no title, 18 October 1951.

While it manufactured explosives, it was also actively looking into new areas of activity [Penrose and Pitelis, 2002, p. 83]. In 1918, the US Department of Agriculture informed Hercules that many farmers and naval stores companies were looking for a new way to extract dead pine stumps from the ground using explosives. In 1918, dead pine wastelands covered 76 million acres (307,561 km^2), and they increased by 10 million acres (40,468 km^2) each year. Yaryan experimented with dynamite but with little success. The explosion could not be too strong; otherwise, the stump would be completely destroyed, and he did not have enough experience with explosives to calibrate them. In 1919, Hercules contacted Yaryan, and the two companies started collaborating. Very rapidly, satisfied with the results, Hercules Powder Company not only started selling its explosives for stump extraction but decided to enter the wood naval stores market on its own and add to its portfolio turpentine, rosin, and pine oil. One of the chemists noted in the company's internal bulletin: "For better or for worse, we embarked on a naval stores venture."[219]

Economic historians have characterised Hercules Powder Company's decision to enter the wood naval stores market as "brave" and surprising, as the company had no previous experience in the field [Dyer and Sicilia, 1989, p. 65]. Hercules developed its own industrial infrastructure but also bought Yaryan's plants in Brunswick and Gulfport. With these moves, Hercules quickly became the largest producer of wood naval stores in the world, controlling around 60 per cent of the global production. It must be reiterated though that in the early 1920s, the wood naval stores production accounted for only 3 per cent of all types of naval stores produced worldwide. It was then a marginal outlet that did not threaten the turpentine and rosin coming from traditional gum resin distillation.

Hercules was not the only one to show interest in this new market. Back in 1913, a certain Newport Company constructed its own plant, but its products were almost entirely used internally [Dyer and

[219] Othmer Library (Philadelphia), Hercules Inc. Series III — Public Relations-1, box 5, Naval Stores History 5/4. Editorial, *Mixer*, March 1920.

Sicilia, 1989, p. 66]. Other major players entered the market a little after Hercules — the Continental Turpentine and Rosin Company (1921), Acme Products (1922), Dixie Products Company (1928), Phoenix Naval Stores Company (1933), and Southern Naval Stores (1936). The market was expanding rapidly, and in 1942, twenty years after Hercules' decision, the annual production of wood naval stores equalled that of the gum naval stores.[220] However, this does not mean that the beginnings were easy.

Just after Hercules acquired Yaryan's company, the industry faced a crisis and the production of wood naval stores in Hercules' plants fell by 50 per cent.[221] However, the major problem with naval stores extracted from pine wood remained their poor quality. Wood rosin was ruby red, sometimes almost black, having a tendency to crystallise and smell bad. Potential outlets included inks, linoleum, and road coating, but overall, about 70 per cent of the market was unavailable for this type of rosin. Wood turpentine certainly had fewer issues but was also considered less well-defined and inferior to that of gum turpentine. And yet, the directorship took a firm decision to continue investing in its scientific laboratories to improve these products in spite of temporary losses. As previously explained, the gum naval stores industry had only one major laboratory — the USDA Bureau of Chemistry (divided between Washington and Olustee). Hercules had dozens of chemists spread across tens of small plant laboratories. It achieved a critical mass necessary to create a specialised field of "wood naval stores chemistry", developed almost entirely by the private sector.

Between 1921 and 1927, the improvement of wood distillation methods increased the rosin yield by 40 per cent. In 1928, Hercules established a full-fledged research and development section exclusively devoted to the development of new resin-based products. The same year, the Hercules Powder Company introduced pale rosins

[220] Othmer Library (Philadelphia), Hercules Inc. Series IV — Hercules History 2, Box 5. Anonymous, "Naval Stores History We Never Knew 'til Now".

[221] Othmer Library (Philadelphia), Hercules Inc. Series IV — Hercules History 2, file "Naval Stores". "History of Hercules Naval Stores Business", 24 September 1945.

on the market, comparable to the ones coming from gum distillation, making their way to the soap trade and many others. The company also developed more new methods aiming to extract wood turpentine and terpenes, such as di-pentene and α-terpineol (the major component of pine oil). The company's portfolio continued to diversify.

Why did Hercules' directors pursue these lines of activity so relentlessly? The answer lies in the fact that from the outset of the industry, they identified persistent weaknesses in the existing gum naval stores industry that were hard to overcome but avoidable if a different manufacturing had been adopted. Already in 1921, one of Yaryan's and then Hercules' engineers explained that he was convinced of the enterprise's long-term success.

> Gum rosin and gum turpentine, the original and still the principal products of the naval stores' industry, have been and still are, produced under such primitive conditions and by such methods, as to preclude any material change or improvement in these products, or the production of products of a uniform character ... [They] have therefore been generally accepted as standards of quality, without recognition of the fact that products of higher quality and greater uniformity could be produced under different conditions, with improved methods and proper control [Lockwood, 1921, p. 247].

The wood naval stores fulfilled these hopes, and the success of the new industry was unquestionable. In the late 1930s, wood rosins became more sought after than gum rosins due to the fact that their quality was much more uniform. Varying meteorological conditions and gum harvesting methods were no challenge to the wood naval stores industry, and the industrialisation of the production in huge plants made uniformity much easier to achieve for wood naval stores producers than for little scattered gum distillation workshops. Also, the gum naval stores industry was seasonal, while the wood resin industry delivered products on a regular basis all year long. Finally, the entire production process enabled the specialisation of production lines, facilitating the manufacture of more sophisticated products.

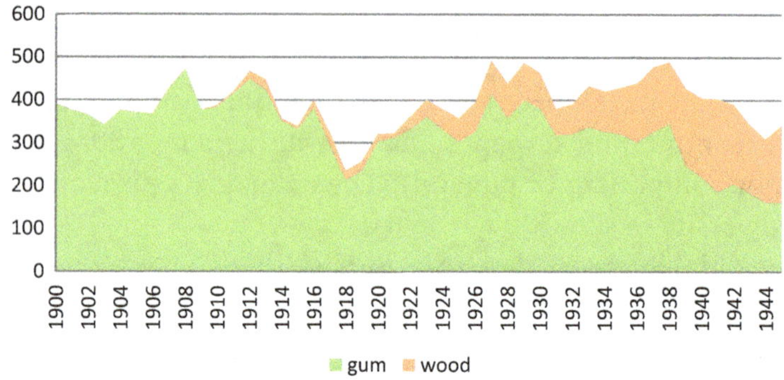

Figure 52: Rosin production in the United States by origin (thousands of tons).[222]

Figure 52 shows that in the 1920s, the production of wood rosin was a tiny fraction of the total global rosin production, but by the end of the war, the US's production of wood rosin equalled that of gum. The rate of growth was staggering.

On the whole, wood rosins reshaped the US rosin market. Ironically, this success was possible thanks to the Pine Institute in France. In one Hercules internal report from 1939, the authors cited, for example, Dupont and Barraud as leading references in the field [Borglin, 1939]. In fact, the Americans, by the end of the 1930s, still lagged behind in terms of fundamental research. Both Hercules Powder Company and the USDA were, above all, focused on practical improvements and had to rely on French publications on the composition and properties of rosins. In a certain way, the knowledge of the French scientists was used against them. In 1946, one company wrote to Brus about certain uses of the wood rosins they imported from the US. Brus responded abruptly that there was an equivalent product manufactured in France and that the company should buy from the Landes producers instead of importing from abroad.[223] It shows, however, that

[222] Established from data in [Zinkel, 1989, p. 61]; and in *Naval Stores International Yearbook* (Naval Stores Review, New Orleans), 1948.

[223] Departmental Archives of Gironde, Pine Institute Archives, file "Correspondance 1946, M". Letter from Manufacture des Lièges Ouvrés & Isothermiques to the Pine Institute, 25 November 1946.

the American wood naval stores companies were actively promoting their products on the European markets. The Pine Institute researchers were themselves impressed by the diversity of products that Hercules could extract from wood rosins (brands such as Petrex, Lewisol, Vinsol or Pentalyn).[224] The rapid rise of the wood naval stores industry may give the impression that the gum naval stores stood no chance. Yet, the traditional gum resin producers did not remain passive and tried to carve out their own niche with the help of the USDA.

4. Gum Naval Stores During and After World War II

During a significant portion of the early twentieth century, the American gum resin industry was massive but lagged significantly behind its French counterpart in terms of technological advancements. And yet, after World War II, the situation reversed. The American naval stores industry turned into a beacon of progress, setting the standard for other countries to follow, including France, which had to suddenly catch up with its main competitor. This power shift was, however, a tumultuous and complex process conditioned by numerous factors. Not only that, despite tremendous progress over the period, the industry had to face inevitable decline in the 1960s, just like its French counterpart.

The Changing Face of the American Naval Stores Industry (mid 1930s — mid 1950s)

Four major shifts took place between the 1930s and the 1950s, profoundly transforming the trajectory of the American resin industry (Table 11). The first shift pertained to how decision-makers perceived the naval stores industry, the second to the technical state of the industry, the third to the organisation of scientific research, and the fourth to the overall industry structure. To comprehend how the

[224] Departmental Archives of Gironde, Pine Institute Archives, file "Correspondance 1949, H". Letter from P. Legendre to Hercules Powder Company, 4 May 1949.

Table 11: Four shifts in the gum naval stores industry between the 1930s and the 1950s

Perception Shift	Naval stores became viewed as a tool of conservation instead of a threat to forests
Technological Shift	Military requirements and war-induced relations with the wood naval stores industry led to the transfer of technology to the gum naval stores industry
Scientific Organisation Shift	Concentration of naval stores research in the South (thanks to Utilisation Research) and then in Olustee itself
Industrial Organisation Shift	Multiplication of turpentine farmers and the reduction of turpentine distilleries

Americans successfully caught up with the French in terms of overall development, the following paragraphs will discuss each of these shifts in turn.

First, the social imaginary surrounding the gum industry, as well as the rhetoric justifying the resin research funding, changed. In the 1910s and the 1920s, the naval stores business was considered a destructive force that ravaged the southern forests and threatened their survival — conservationist scholars compared turpentine operators to voracious locusts. In 1920, a US Senate report predicted that forests would disappear altogether if the current usage patterns were maintained. However, with the rise of increasingly sustainable conservation practices, also due to the disastrous Dust Bowl, the role of resin extraction in southern forests was reassessed. Over the next few years, the perception of the gum naval stores industry turned 180 degrees. In 1937, the industry was called "the economic key to successful reforestation of the South".[225] The naval stores industry was now seen as a factor that would allow the forests to survive

[225] US National Archives (College Park), Records of the Bureau of Agricultural and Industrial Chemistry, entry 32, ARC ID# 568557, General Correspondence 1935–1939, box 26, file "J. E. Lockwood, Naval Stores". Letter from C. Speh to J. E. Lockwood, 27 April 1937.

because it brought profit as long as the forest was there, thus encouraging replantation. In the middle of the 1930s, Veitch marvelled over the "unbelievable come-back of pine trees", especially in Georgia, and all this was driven by the economic interest of the naval stores industry.[226]

The second major shift in the American gum naval stores industry was technological in nature and due to World War II. In occupied France, while the overall impact of the war was ambiguous, as explained in Chapter 3, one thing was clear: the resin industry could not develop in service of the national defence. In the US, on the contrary, gum resins played an important role in the military effort by providing a range of indispensable materials, *e.g.*, turpentine was used in the manufacture of camphor (imported before from Germany and Japan) necessary for smokeless powder. Another component synthesised from turpentine was isoprene for the production of rubbers (neoprenes and bunas). Pine oil, a byproduct of the wood resin industry made of terpineol, was used for mineral flotation and in the manufacture of cheap disinfectants and detergents massively consumed by the army. The copper resinate was used for preventing mould growth. Finally, the war was also a golden age for the manufacture of cheap rosin soaps and particularly all sorts of paints and varnishes for tanks, grenades, helmets, munitions, and so on.[227]

The industry's importance to the army was explicitly recognised by the US government. On 16 January 1942, President Franklin Delano Roosevelt established the War Production Board (WPB) to coordinate cooperation between the government and the industry. The WPB had 85 industrial committees, and one of them, section 27, worked on protective coatings, paints and varnishes. This section had several smaller subcommittees, including one devoted

[226] US National Archives (College Park), Records of the Bureau of Agricultural and Industrial Chemistry, entry 69, ARC ID# 3887383, Subject File Relating to Naval Stores Research, 1903–1935, 00530: 052/024/00, box 4, file "Memoranda 1927 — June 30, 1935". "Program on Fundamental Research on Naval Stores", 1935.

[227] US National Archives (Atlanta), RG310, Agricultural Research Service, Publication Records Set, 1907–1939, Record Copy of Station Publications 401–463. R. Palkin, "Turpentine in the war effort".

specifically to wood naval stores and another one to gum naval stores. Their meetings were, however, often held together with the participation of Veitch and later Speh (who replaced Veitch as the USDA's naval stores laboratory director). Discussions concerned the application of final products, yearly production, and the general market situation. Most importantly, they also provided a formalised space for exchanges between the two communities, fostering the exchange of ideas. For the more primitive agricultural gum naval stores industry, this encounter with its industrialised and rapidly developing competitor was particularly revealing. The know-how stemming from these exchanges, as well as from the contact with the army itself, was essential for further modernisation of the entire sector.

The third major shift in the gum naval stores industry concerned the organisation of its scientific infrastructure. In 1938, the US Congress authorised the construction of four huge regional agricultural laboratories in order to foster the so-called utilisation research. They were situated in Albany (California), Peoria (Illinois), Philadelphia (Pennsylvania), and in New Orleans (Louisiana) [Finlay, 1990]. The last one, the Southern Regional Research Laboratory, worked initially on cotton, but in 1941, the Washington Naval Stores Research Division moved into its premises.[228] The New Orleans laboratory continued its work on both fundamental and technological problems. In parallel, fourteen chemists and technicians continued their work in Olustee. This geographical rapprochement fostered collaboration between the two units, which eventually merged in 1951 when the New Orleans personnel were simply transferred to Olustee. In total, the group that year consisted of 24 employees.[229] They were

[228] US National Archives (Atlanta), RG310 Agricultural Research Service, Publication Records Set, 1907–1939, Record Copy of Station Publications, 578–653. E. L. Patton, "Research on Gum Naval Stores — New Information and New products", *AT-FA Journal*, May 1957.

[229] US National Archives (Atlanta), RG310, Agricultural Research Service, Publication Records Set, 1907–1939, Record Copy of Station Publications, 501–577. Editorial, *Florida Section American Chemical Society Journal*, February 1952.

directed by an engineer and war veteran, Colonel E.L. Patton.[230] In general, the Olustee laboratory was fully comparable to the Pine Institute over the period, if not bigger (Ph.D. students excluded).

Finally, the last great shift of the 1930s and 1940s concerned the organisation of the gum resin industry itself — the number of turpentine operators increased from 300 in 1930 to 40,000 in 1949. Turpentining became a secondary activity for thousands of small farmers, as well as for private forest owners for whom the paper industry was the first outlet. The era of major turpentine operators who borrowed the land to exploit it exclusively for turpentine came to an end [Shingler, 1948, p. 104]. Veitch rejoiced at this evolution: "I have always hoped that the naval stores industry would develop along this line rather than be controlled by a few big producers."[231] While the number of gum providers was growing exponentially, distillation underwent an opposite trend. In 1930, the gum naval stores industry had around 1,000 distillation workshops scattered across the forest. In 1949, there were no more than 28 "central plants". This concentration enabled a better level of standardisation as well as the construction of expensive installation that improved the quality of the products [Patton and McKennon, 1950, p. 78].

Visitors from France: The Trip of the French Resin Industry's Delegation to America

It was precisely after World War II, when the four shifts were almost complete, that Legendre, the technical director of the Pine Institute, alongside U.C.R. members representing the Landes industry, decided

[230] Patton started his job at the laboratory already back in 1933 but he was mobilised in 1941. He came back to the bureau in 1946. See: US National Archives (Atlanta), RG310 Agricultural Research Service, Publication Records Set, 1907–1939, Record Copy of Station Publications, 464–500. "E. L. Patton Heads Naval Stores Research".

[231] US National Archives (College Park), Records of the Bureau of Agricultural and Industrial Chemistry, entry 32, ARC ID# 568557, General Correspondence 1935–1939, box 26, file "Naval Stores — Miscellaneous". Letter from F. P. Veitch to Brown, 22 March 1937.

to visit the US to discover the state of the American scientific and industrial infrastructure. Only 15 years before, it had been the French industry that inspired the Americans, who painstakingly adopted the techniques that had been known in France for many years (*e.g.*, resin cups, resin cleaning, steam and continuous distillation). In 1947, the situation changed drastically, and it was the French who had to suddenly confront a new reality and try to learn from the Americans.

The journey took two months and was very intense. The French started in New York and then headed to Washington, Savannah, Brunswick, Jacksonville, Lake City, Pensacola, New Orleans, Chicago, and Madison before returning to New York (Figure 53). They visited the most important industrial and scientific centres of gum and wood naval stores industries. Educational value aside, the trip was also an important cultural experience. Legendre wrote to Brus that "America is a surprising country — admirable and disappointing at the same time" and noted that American food was "tasteless" and vastly inferior to French food.[232]

In Washington, Legendre encountered Speh. Later, he visited the Olustee Station, where he talked with Shingler and Patton, and then the New Orleans lab, as well as a forestry lab in Madison. He found the American laboratories to be "splendid" and the salaries of federal researchers to be "pretty nice".[233] He underlined the importance of private sector research, notably in the context of Hercules Powder Company. Legendre met the representatives from Hercules Powder Company in Wilmington, but he deplored the company's engineers for keeping their secrets "jealously" and not wanting to give any details on their techniques of rosin treatment.[234] He particularly regretted that he could not visit the interior of the plants. He also described in his letters

[232] Departmental Archives of Gironde, Pine Institute Archives, box "USA", file "Voyage P. Legendre". Letter from P. Legendre to G. Brus, 18 April 1947.

[233] Departmental Archives of Gironde, Pine Institute Archives, box "USA", file "Voyage P. Legendre". Letter from P. Legendre to G. Brus, 18 April 1947.

[234] Departmental Archives of Gironde, Pine Institute Archives, box "USA", file "Voyage P. Legendre". Letter from P. Legendre to Mme Legendre, 19 March 1947.

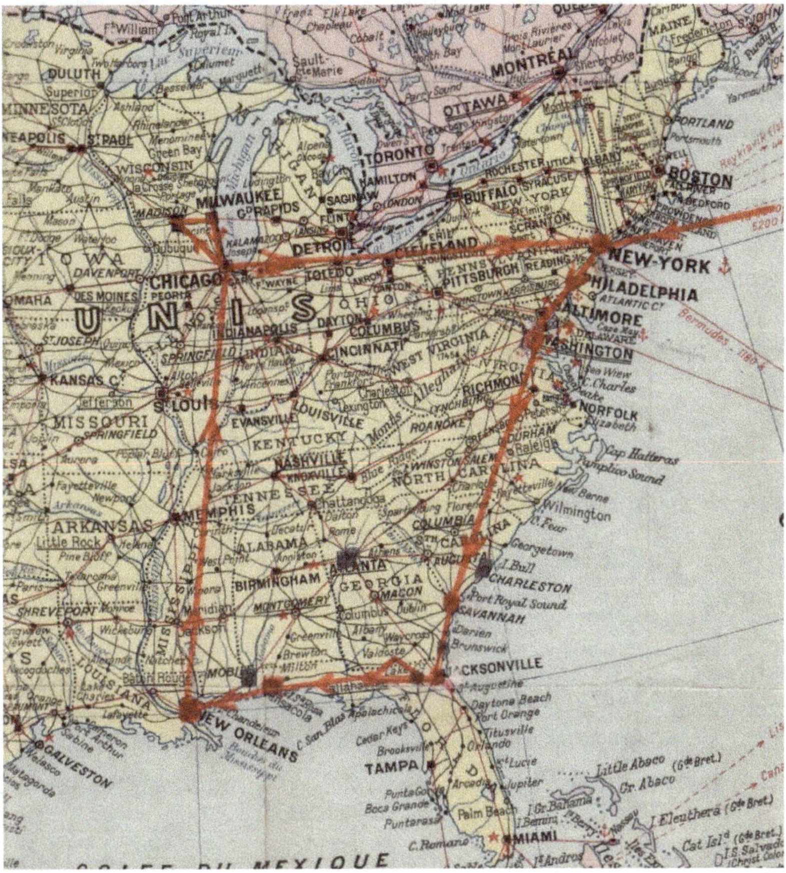

Figure 53: The travel plan established by Pierre Legendre.
Source: Departmental Archives of Gironde, Pine Institute Archives, box "USA", file "Voyage P. Legendre".

to Brus various forestry techniques, such as chemical undergrowth destruction or resin harvesting with acid, which were later introduced in France. In addition, he noted with enthusiasm the beneficial effects of the concentration of distilleries, a policy that the Pine Institute advocated in France. Finally, he observed that the battle between gum and wood naval stores was underway and collected samples of wood naval stores to analyse them in the Pine Institute on his return.

The trip was covered in the largest professional publication, *Naval Stores Review*, which explained that the French had come to America to revitalise their own resin industry after the war. The French had to suddenly catch up with the industry that had lagged far behind them only two decades before [N. N., 1947, p. 3]. The visit was also mentioned in a local newspaper, *The Valdosta Daily Times*. This article notably mentioned that the French forest workers were paid considerably less than the Americans, even though they remained more productive [Bradwell, 1947], another sign of change in the industry, which had historically relied on underpaid labourers.

American Gum Resin Chemistry After World War II

After the war, the New Orleans and Olustee laboratories tackled a vast range of new scientific problems, some very complex. There was a clear opening up towards fundamental research. Laboratories bought specialised equipment for experiments, such as an infrared spectrometer (to study the oxidation reactions in terpenes and resinic acids), an ultraviolet spectrometer, ozonisers and autoclaves.[235] What were the main lines of research? As noted before, in 1928, Veitch had presented three lines of activities built around the improvement of distilleries, knowledge dissemination, and elaboration of new products. In 1952, when interviewed on the laboratory's missions, he enumerated four major priorities: (1) Assisting the industry, (2) Studying pinic acid, (3) Studying metal resinates, and (4) Studying the composition and properties of gum rosin.[236] The

[235] US National Archives, (Atlanta), RG310, Agricultural Research Service, Publication Records Set, 1907–1939, Record Copy of Station, Publications 501–577. "Research Program for Naval Stores Station", *AT-FA Journal*, April 1955, p. 2.

[236] US National Archives (Atlanta), RG310 Agricultural Research Service, Correspondence and Reports 1952–1956, box 1, file "Acids, Pinic 1952". "Long distance call from C. F. Speh", 6 June 1952.

first and last categories were general, emphasising the laboratory's versatility, while the second and third categories were more specific and much more concrete. Pinic acid, in particular, deserves more attention.

By the end of the 1940s, the American army was not satisfied with cheap mineral lubricants used in the turbo-jet engines in aircraft and servo-mechanisms, such as fire control equipment on naval aircraft. As a result, it began to explore alternative lubricants based on sebacic acid derived from castor oil, specifically 2-ethylhexyl sebacate. What was the advantage of these plant-based lubricants? — "A minimum change in viscosity over a wide temperature range, resistance to oxidation, and permanence of viscosity (no loss of light ends due to evaporation). These latter two features make lifetime lubrication possible in many applications and thereby greatly reduce maintenance costs".[237] The challenge, however, was that castor grains had to be imported, prompting the army to seek out a cheaper domestic alternative. In 1948, it was shown that the n-hexyl of pinic acid from turpentine spirit had similar properties and could potentially serve as a substitute for castor oil. In the early 1950s, several private companies collaborated closely with the Olustee Station on the improvement of pinic acid production methods. A semi-industrial pilot plant was even established for this purpose. Dozens of researchers and technicians worked exclusively on this topic, and federal investments were significant.

Over the same period, an industrial magazine, *Chemical Week*, asked rather sceptically, "Is it worth the trouble?" It questioned the entire investment in pinic acid and pointed out the steady supply of castor beans that the US Army could stock. The USDA's answer was "yes" — pinic acid was not only important for the national defence but also for the entire naval stores industry, which struggled during

[237]US National Archives (Atlanta), RG310 Agricultural Research Service, Correspondence and Reports 1952–1956, box 1, file "Acids, Pinic 1952". E. L. Patton, "Reports on Actions taken to intensify work on pinic acid", 25 November 1952.

the economic crisis. *Chemical Week* also asked about its uses outside the military world. The answer was that if the prices became low enough, many companies might turn to pinic acid for the manufacture of plastifiers and synthetic fibres.[238] Internal reports of the Olustee Station left no doubt: "The field of dibasic acids is very competitive [...]. Interest in pinic acid will be at peak only so long as the supply of sebacic acid is critically short."[239] The Olustee Station collaborated with the Georgia Institute of Technology on improving the overall yield of their process. This outlet must have been promising because Hercules Powder Company, as well as the Polytechnic Institute of Brooklyn, worked on exactly the same question independently.[240] And yet, by 1957, there was still no full-fledged industrial production, and the entire operation remained overly expensive. The Olustee Station did not abandon the topic, and this line of research was continued, especially in the context of potential uses in the fabrication of plastic materials. Nonetheless, the relative importance of the topic decreased as the years went by.[241]

In 1956, the Olustee Station was divided into three research groups: (1) Fundamental Research Unit, (2) Technological Research Unit, and (3) New Products Unit [Woodroof and Patton, 1956, p. 64]. The station was directly supported over the period by the American Turpentine Farmers Association (AFTA), which suggested in 1955 to extend the station, employ more chemists, and explore

---

[238] US National Archives (Atlanta), RG310 Agricultural Research Service, Correspondence and Reports 1952–1956, box 1, file "Acids, Pinic 1952". E. L. Patton, "Reports on Actions taken to intensify work on pinic acid", 25 November 1952.

[239] US National Archives (Atlanta), RG310 Agricultural Research Service, Correspondence and Reports 1952–1956, box 1, file "Acids, Pinic 1952". G. S. Fisher, "Production of Pinic Acid from Turpentine and Evaluation of Derivatives for Defense Uses", 11 December 1952.

[240] US National Archives (Atlanta), RG310 Agricultural Research Service, Correspondence and Reports 1952–1956, box 1, file "Acids, Pinic 1953–1954". Letter from D. Othmer to F. Bellinger, 16 October 1953.

[241] US National Archives (Atlanta), RG310 Agricultural Research Service, Correspondence and Reports 1952–1956, box 1, file "Gum naval Stores Industry Committee, 1957". Report on the Meeting of the Gum Naval Stores Industry Research Committee, 9 May 1957.

other promising studies.[242] In many ways, the ATFA played a similar role to the U.C.R. in France and its pleas must have had some effect on the station's funding because the station developed a new ambitious strategy in 1957. Pinic acid aside, Olustee was to study (1) rosin neutralisation (a topic that occupied the French chemists as well), (2) non-acid portion of rosins (around 10 per cent by weight), (3) a new type of varnish from formaldehyde-based rosins, (4) the use of rosin derivatives for plasticisers, (5) detailed composition of gum turpentine, and (6) oxidation of terpenes.[243]

In spite of the quality of these studies, one element in the internal reports was striking. One of the topics studied in Olustee in 1960/1961 was the characterisation of non-identified compounds in resin. The chemists were led to a surprising discovery that the proportions of different chemical compounds in the gum varied not only between different species but also between pine trees of the same species in the same region. They indicated that the proportion of α-pinene could vary from 45 to 85 per cent, β-pinene from 13 to 42 per cent, and β-phellandrene from 0 to 20 per cent in longleaf pine turpentines. If these findings surprised the American chemists in the 1960s, it meant that they had earlier studied the commercial mixture available on the market but not the resin coming directly from the trees. These results had already been well known in France during the 1920s, notably (but not exclusively) thanks to the works of Marcelle Barraud. Her publications were mentioned in American sources of the period, indicating that, at least to some extent, this knowledge must have passed to the US. However, the general disinterest in theoretical problems among Americans in the 1920s, coupled with rapid modifications of research organisations and of personnel, led to a situation in which Barraud's

[242] US National Archives (Atlanta), RG310 Agricultural Research Service, Correspondence and Reports 1952–1956, box 1, file "Gum naval Stores Industry Committee, 1957". Notice from 15 December 1955.

[243] US National Archives (Atlanta), RG310 Agricultural Research Service, Correspondence and Reports 1952–1956, box 1, file "Gum naval Stores Industry Committee, 1957". "Report on the Meeting of the Gum Naval Stores Industry Research Committee", 9 May 1957.

works were simply forgotten. The younger generation of chemists in the 1960s were not familiar with older works written in French that came from a region whose glory days were long gone.[244] After the war, the Pine Institute of Bordeaux was almost absent in the pages of the *Naval Stores Review*, and the French resin chemistry school was slowly becoming irrelevant.

When compared, the French and American resin-related research infrastructures after World War II shared many similar characteristics. The Pine Institute had a relatively comparable size to the Olustee Station, and the producers' associations, the N.U.C.R. and the ATFA, respectively, influenced the general direction of studies in a similar manner. Some characteristics were clearly different; for example, the Olustee Station, unlike the Pine Institute, never studied resin harvesting itself. This line of research remained in the purview of the USDA Forest Service. Not only that, whereas in France, the discipline was centralised in Bordeaux; in post-war America, Olustee aside, more and more universities across the country were becoming interested in the field; for example, the University of Florida had no less than three laboratories working on naval stores [Hawkins, 1948, p. 62]. There were also researchers in Georgia Institute of Technology interested in gum naval stores in the late 1960s [Tze Chiang *et al.*, 1968]. Overall, the American naval stores research panorama, with all the federal, academic, and private initiatives, and with two competing industries of gum and wood, was considerably larger and richer than the French one, which was dominated exclusively by the Pine Institute.

The End of (Gum) Naval Stores

Despite all the investment in its reconstruction, the gum naval stores industry started losing ground to the wood naval stores industry in the 1950s. Between 1950 and 1959, the gum production in the US decreased by 50 per cent, and the number of major

[244] US National Archives (Atlanta), RG310 Agricultural Research Service, Correspondence and Reports 1952–1956, box 3. "Work Project Annual Report, July 1960 — June 1961".

distillation plants fell from 26 to 19, this time not due to a concentration policy but because of bankruptcies.[245] Over the period, the famous resin market in Savannah also disappeared, and it was replaced by a new entity: the Naval Stores Market News Office. It has to be noted that the Savannah market was dissolved after an investigation by the USDA, which discovered that some traders had manipulated resin prices.

And yet, researchers held onto their optimism. In 1952, in an article titled "Best Days Ahead for the Naval Stores Industry", Speh argued that even if the situation of the naval stores had been gloomy in the 1930s, the hope was sparked anew "not because of [trade] statistics but because of research".[246] It was science that was supposed to save the industry yet again in the 1950s. In 1957, the Olustee's director, Patton, affirmed that one should not lose hope and that better days would return:

> In 1938 there was no Nylon ... no television industry, and in the U.S. no synthetic rubber industry. No one knew how to split an atom; there was no jet aircraft, guided missile, U.S. helicopter, "mechanical brain" ... it can be assumed with confidence that the total effort in research laboratories all over the world will continue to bring spectacular advances and new things that will add to our good health, comfort, pleasure and leisure. We hope this will be the case in naval stores research ... Research will give us the knowledge needed to extend and improve this growth period.[247]

---

[245] Georgia Historical Society (Savannah), Naval Stores Collection, M 2022, Series 1–1, folder 3 "History of the U.S. Department of Agriculture". "The Gum Naval Stores Loan Program", 1 April 1960.

[246] US National Archives (Atlanta), RG310 Agricultural Research Service, Publication Records Set, 1907–1939, Record Copy of Station Publications, 501–577. C. Speh, "Best Days Ahead for the Naval Stores Industry", *AT-FA Journal*, September 1952.

[247] US National Archives (Atlanta), RG310 Agricultural Research Service, Publication Records Set, 1907–1939, Record Copy of Station Publications, 578–653. E. L. Patton, "Research on Gum Naval Stores — New Information and New products", *AT-FA Journal*, May 1957, p. 2.

In a certain way, Patton was right. Progress in the naval stores industry brought fruit but not in favour of the gum naval stores. One of the arguments frequently used against the wood naval stores was that the supply of old stumps was not inexhaustible. From its inception, the wood naval stores industry relied on non-renewable materials, but thanks to conservation practices, surfaces of wastelands covered with dead pine stumps were decreasing. In 1948, USDA researchers believed that, due to an inevitable scarcity of raw materials in the wood naval stores industry, the gum naval stores industry would soon come back to its former glory.[248] But Hercules chemists were fully aware of the challenge. They notably worked on the process of artificial stump "resinification". Normally, in order for a stump to become rich in resin, it had to stay in the ground for a period ranging from 10 to 25 years. However, these stumps can be treated with various acids to force resin generation and reduce this period more than by half. The wood naval stores industry was trying to become more renewable and better integrated into a more sustainable forest economy [Zinkel, 1989, p. 144].

However, both the wood and gum industries encountered two significant obstacles. First, petroleum derivatives became ubiquitous, with white spirit rapidly gaining prominence and dominating the paint and varnish industry.[249] Second, and more importantly, the paper industry became a major produce of naval stores in the 1940s. During the manufacture of paper sizing in the Kraft process (that entails the treatment of wood chips with sodium hydroxide and sodium sulfide), a byproduct known as tall oil was generated — a complex mixture with a huge proportion of fatty and resin acids. On top of that, the paper mills could easily manufacture the so-called

[248] Georgia Historical Society (Savannah), Naval Stores Collection, M 2022, Series 9, file "Publications — articles and Newsclippings 1930–1950". E. L. Patton, Naval Stores, June 1948.

[249] National Archives (Atlanta), RG310 Agricultural Research Service, Correspondence and Reports 1952–1956, box 1, file "Acids, Pinic 1952". "The questions prepared by Mr. John Kend, McGraw-Hill Publications, on synthetic lubricants from turpentine", *Chemical Week*, no date.

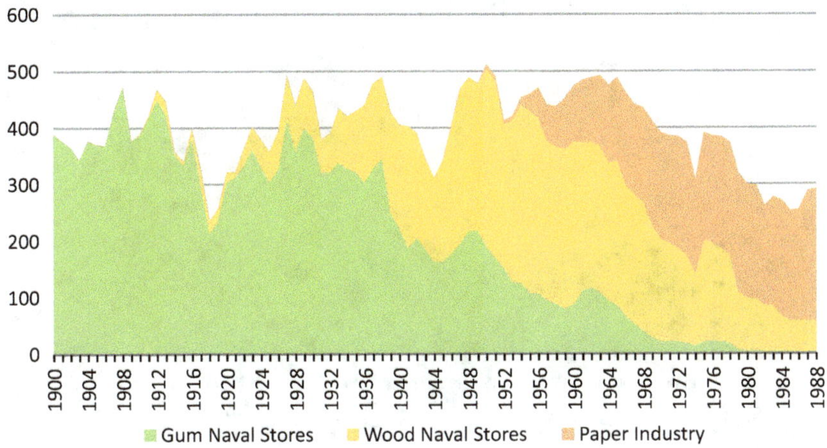

Figure 54: Rosin production in the United States (in thousands of tons), 1900–1988.

sulfate turpentine. Papermakers realised that these low-quality byproducts could be cleaned and improved to compete with the wood and gum naval stores. In fact, they "became" wood naval stores by default. Figures 54 and 55 illustrate the dazzling growth of this sector.

While wood rosin was dominant in the market during the immediate post-war era, wood turpentine failed to attain the same status and was swiftly overtaken by sulfate turpentine from the paper mills. The success of tall oil and sulfate turpentine can be explained by their cost advantage — they were generally around 10 per cent cheaper than their competitors. This cost advantage stemmed from the fact that these byproducts were generated during paper production, allowing producers to accept lower profit margins.[250] They also had all the advantages of the wood naval stores (industrialised production, no variability throughout the year).

[250]US National Archives (Atlanta), RG310 Agricultural Research Service, Correspondence and Reports 1952–1956, box 2. "Foreign Production and utilization as a factor in the domestic supply of naval stores", Naval Stores Work Conference, 1956.

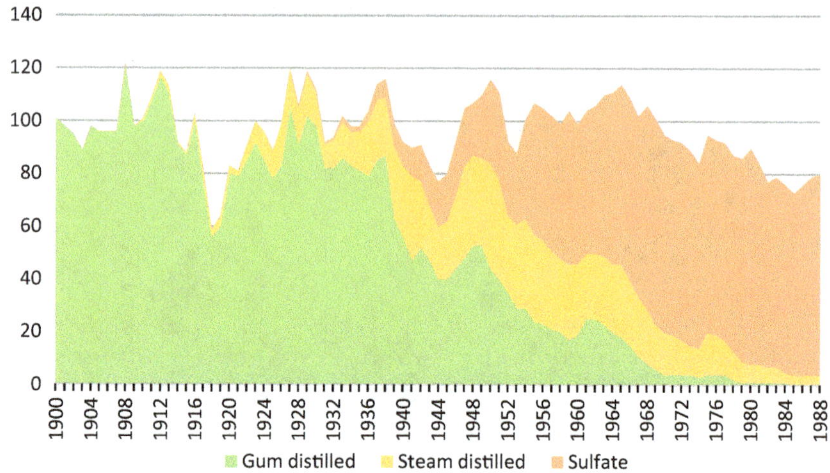

Figure 55: Turpentine production in the United States (in thousands of tons), 1900–1988.

In the US, the gum naval stores industry came to a definitive end in the 1970s, coinciding with the fate of the French industry during the same period. Consequently, the closure of the gum naval stores industry also marked the disappearance of gum resin chemistry. The Olustee Station shut down in 1973, just four years after the Pine Institute nearly faced a similar fate. However, the Pine Institute managed to avoid closure by diversifying its involvement in the lumber and paper industries.

In the second half of the twentieth century, the connection between the modern resin industry and its forerunners began to dissipate. This shift is particularly visible in the vocabulary used by the industry's stakeholders. Already in the 1950s, the Glidden company changed the name of its Naval Stores Division to Southern Chemicals Division.[251] In the decades that followed, the Tall Oil Association became the Pulp Chemicals Association. The

[251] Georgia Historical Society (Savannah), Naval Stores Collection, M 2022, Series 1–1, folder 3 "History of the U.S. Department of Agriculture". M. S. Briggs, "Naval Stores Stocks and Distribution", 6 October 1956.

most symbolic change occurred in 1998 when the oldest publication in the field, *Naval Stores Review*, changed its name to the *Pine Chemicals Review*, which is still published today [N. N., 1998]. With this rebranding, the history of the naval stores in the US came to an end.

Parenthesis on Iberian Peninsula and Eastern Europe

It should be clear by now that the histories of the French and American naval stores industries are intertwined. France and the US were the only two countries where resin chemistry could develop as an autonomous discipline. However, the French influence extended much farther to the Iberian Peninsula and Eastern Europe. It is beyond the scope of this book to explore the history of all these respective industries, but identifying the key knowledge flows between them and France is necessary to understand the global context in which both the French and American research infrastructures developed.

In Spain, the resin industry was most likely established by a group of Franco-Spanish entrepreneurs who founded a resin company in Segovia in 1862. The company not only modernised distillation techniques, it also introduced Hugues' pots to Spain, which became mandatory there as early as the 1860s. The industry flourished during the American Civil War, like the French one, but it barely survived the period that followed. Although the Iberian Peninsula had vast pine forests, they were scattered throughout the territory and were less concentrated than the Landes Forest near the major port of Bordeaux. Thus, the exploitation and transportation of end products were much more expensive. Additionally, a commercial treaty between France and Spain from 1879 lowered taxes on resinous products exported from France to Spain. The maritime ports in Bilbao and Santander (but also important trade centres such as Sevilla) were flooded by French resins, which were much cheaper to transport by sea from Bordeaux than by land from the centre of Spain [Moral, 1921, p. 191; Delgado, 2016].

In the last decade of the nineteenth century, the Spanish resin industry was saved by a Spanish Resin Union, a form of trade trust created in 1898. The union controlled around 70 per cent of domestic production in the first years of the twentieth century. In 1928, in order to improve the industry technologically, the Spanish established a special section devoted to resins in their forest service.[252] The service was open not only to forestry but also to resin chemistry, and was envisaged to create a publication similar to *Bulletin de l'Institut du Pin* or *Pine Institute of America Abstracts*, whose successes it wanted to emulate. The importance of the Spanish scientific research initiated in 1928 continued to grow. Brus' correspondence suggested that the relations between the Spanish scientists and the Pine Institute were cordial.[253] While the French occasionally used the Spanish expertise, no structure similar to the Pine Institute or the Olustee Station was ever created in the country, and even in 1962, young Spanish students were trying to gain admission to the Pine Institute as interns.[254]

As for Portugal, the techno-scientific lag was even more pronounced and Portuguese methods remained relatively primitive and destructive for much longer than in Spain. In fact, in the 1910s, the Portuguese even employed a number of French forest workers so they could impart their French methods to the local labour [Poudenx, 1947]. Unlike Spain, the Portuguese government did not create any scientific division devoted to resins. The Lusitanian scientists regularly contacted the Pine Institute for analyses and were probably the most visible foreign group among the Institute's contacts. Even major research institutions, such as *Instituto Superior Técnico Química Orgânica* (Higher Technical Institute of Organic Chemistry)

[252] "The Organisation of the Section of Resins", Servicio Forestal de Investigaciones y Experiencas, n. 2, 1928, extracted from *Pine Institute Abstracts*, 3 (9), 1929, p. 10.

[253] Departmental Archives of Gironde, Pine Institute Archives, file "Correspondance 1950, L". Letter from Leun & Cie, 9 January 1950.

[254] Departmental Archives of Gironde, Pine Institute Archives, file "Correspondance 1962, Q-R". Letter from Ministerio de Agricultura, Instituto Forestal de Investigaciones y Experiencas, 11 July 1962.

in Lisbon, contacted the Institute for rosin analyses, indicating a lack of expertise and equipment for such tests.[255]

The scientific and technological lag did not stop the development of the resin industry in the Iberian Peninsula. In 1927, Spain manufactured 5 per cent of the global supply of resin, while Portugal did 1.6 per cent (France produced 22 per cent and the US 67 per cent) [Dupont and Barraud, 1927]. However, by the middle of the 1930s, the collective yield of both Spain and Portugal exceeded France, with Portugal becoming the second largest producer in Europe (and third in the world). By 1960, Portugal was the largest European producer [Zinkel, 1989, p. 61].

As for Eastern Europe, the French connection was less evident. The Russian chemist Dimitri Mendeleev had written back in 1897 that Russian turpentine could play the same economic role as the French or American one.[256] Some attempts were made to introduce turpentine farming in Russia in the 1890s but with little success. It was a certain L.L. Volkov, a Russian forester, who first introduced the gum harvesting technique in the Russian empire after an extended stay in the Landes Forest.[257] In 1908, he started forest experiments near the Polish town of Skierniewice, and soon after, the first resin company was opened in another Polish city of Radomsko. After the liberation of Poland in 1918, these initiatives laid the foundations for the Polish resin industry. However, Volkov moved to Siberia, where he also successfully established new manufacturing plants.

Interestingly, after the Revolution, turpentining came under critique. Some Soviet scientists considered turpentining to be detrimental

---

[255] Departmental Archives of Gironde, Pine Institute Archives, box "correspondance 1958–1959", letter L. Letter from Instituto Superior Technico Quimica Organica, 14 February 1958.

[256] Georgia Historical Society (Savannah), Naval Stores Collection, M 2022, Series 9–1. D. Mendeleev, Tolkovyj tarif ili issledovanie o razvitii promyshlennosti Rossii v svjazi s ee obwim tamozhennym tarifom 1891 g. [A sensible tariff or research about development of the industry of Russia in connection with its blanket customs tariff of 1891] cited in F. A. Mednikov, Turpentining, no date, p. 3,

[257] Georgia Historical Society (Savannah), Naval Stores Collection, M 2022, Series 9–1. F. A. Mednikov, Turpentining, no date, p. 6.

to wood and economically not viable. F.A. Mednikov, another promoter of the turpentining in the Soviet Union, had to defend it using a particular rhetoric. He notably wrote that, according to Dupont, the *Pinus sylvestris* from Eastern Europe was unsuitable for oleoresin extraction. Mednikov accused Dupont of lying in order to protect the interests of "capitalist wood producers", who were afraid of losing huge markets if the Soviet Union started manufacturing its own turpentine and rosin.[258] Yet Mednikov himself recognised some problems, noting that the Soviet Union climate was not favourable for such an enterprise (resin flows better when it is warm) and that the vast Siberian forests were poorly integrated into the Soviet communication network. Little is known about the influence of the French (or American) techniques and scientific discoveries in Eastern Europe, but in Poland and the Soviet Union, there were novel and interesting studies on resin properties and exploitation (for example, see [Grochowski, 1965, pp. 432–447]).

Conclusions: France Dethroned

It may have been the German-speaking chemists, such as Alexander Tschirch, who laid the foundations of resin chemistry in the first decades of the twentieth century, but it was the French who built the discipline, properly speaking, and who constructed the "scientific" or "rational" resin industry. Even before the Pine Institute era, the French forest workers, foresters, inventors, and engineers from the Landes and Gironde developed new ways of forest management and resin extraction in an organic manner. The French system inspired the Spanish, Portuguese, Russians, and many other nations. The Landes industry, along with its scientific infrastructure, became a model to follow for the entire world. The Americans were the most loyal disciples of the French, adapting their technological and organisational solutions as well as emulating their

[258] Georgia Historical Society (Savannah), Naval Stores Collection, M 2022, Series 9–1. F. A. Mednikov, *Turpentining*, no date, p. 15.

institutions. After World War II, however, it was the French who had to catch up with their American colleagues and struggled to stay relevant in the scientific and industrial world of naval stores.

The shift in global influence regarding resin production raises important questions. Is it right to attribute the French decline to a failure in their approach? Paradoxically, it was the Americans who had a laboratory funded by the federal government with public funding, while the French organised their own grassroots institution. Furthermore, despite the French forests and industry being considerably smaller than their American counterparts, it was the French who spearheaded the developments in the field and maintained better-equipped laboratories until World War II. Additionally, when it comes to basic research and theoretical contributions, the authority of French scientists remained unquestionable for a much longer duration.

If we exclusively compare the gum naval stores industry and chemistry (excluding that of wood), the gap between France and the US probably narrowed in the years following the war. The Olustee Station and the Pine Institute remained similar in size for much longer, and the French introduced many of the American solutions (acid spray, concentration of workshops) in their own country. It was, however, the wood naval stores industry that shifted the balance. No entity in France could compete with Hercules Powder Company. The paradox stems from the fact that the American success came precisely from their destructive unsustainable practices over decades, where they transformed thousands of hectares of primal forest into barren wastelands that Hercules could operate for such a long time using dead pine stumps. No such possibility existed in France, where the forest, as artificial and industrialised as it might have been, was exploited in a fully sustainable way. This also had an impact on the production of scientific expertise. The American chemists could circulate between the two industries. The failure of the gum industry did not mean that naval stores chemistry was over; after the Olustee Station was closed, it was simply practised elsewhere, namely, in wood and pulp naval stores businesses. In France, resin chemistry was no more after the 1970s.

Overall, the resilience of the French structures remained impressive, considering the much less favourable conditions it had to face. From the symbolic point of view, the Pine Institute was considered a key authority in the field of resin chemistry in the 1950s, notably for the standardisation of resin products discussed in the following chapter.

Chapter 6

OF LAWS AND RESINS: THE PINE INSTITUTE AND STANDARDISATION

> "*La science ressemble aux chars de ces divinités indoues qui écrasent les fidèles sur leur passage.*"
> André Bodin (1927). Les résineux devant le progrès scientifique, *Bois et Résineux*, 22 May 1927.

The Pine Institute and the American laboratories were not only research institutions but also providers of expertise. They defined the characteristics of properly distilled resin products and established baseline standards that these substances should satisfy on the market. These standards were held in high regard because of their "scientific" character. However, the rich literature on the sociology of standardisation points out the fact that even the standards based on the expertise of scientific organisations are never unequivocally objective, in the sense that they always convey a range of assumptions and linguistics conventions and quite often reflect some type of interests [Egyedi, 1996; Timmermans and Epstein, 2010; Jakobs, 2011]. Standards always result from multifaceted historical processes that are deeply rooted in local contexts. At the same time, every attempt at standardisation involves disregarding social and cultural factors to emphasise neutrality. On the one hand, terminology from the industry is integrated into laboratory practices, creating a common language between the laboratory and the factory. On the other hand, scientists refine these categories by delivering standards that the industry later uses. As specialists in the philosophy of standardisation put it, standards are "both prerequisites for scientific practice and outcomes of scientific expertise" [Huber and Schlaudt, 2015, p. 3].

The goal of this chapter is to address the relationship between scientific expertise and the standardisation of complex chemical substances in the case of pine chemistry. The argument put forth is that the process of renegotiating the definitions of commodities and incorporating them into the categories of modernity did not end in the nineteenth century but remains an ongoing issue. This chapter aims to demonstrate how seemingly neutral objects, such as rosin, rosin soap, and especially turpentine, have been objects of heated debates and how the process of defining them by scientific institutions was marked by extra-scientific considerations.[259]

1. Standardisation of Secondary Resin Products: Soaps and Rosins

The history of resin soaps and rosin colour grades reveals two different roles of the Pine Institute in the process of standardisation. With respect to rosin, the Institute acted as an "active translator" of American norms, adapting them to the French environment, using its scientific authority to legitimise the standards in the local community. In the case of soaps, the Institute mainly provided a forum for discussion, and its bulletin showcased the state of debates and conducted some modest lobbying efforts.

Marseille Soap 72% Extra Pure and Pine Resin

Soap is a combination of a fatty acid and a strong base, such as sodium hydroxide or potassium hydroxide. The fatty acid portion

[259] Some preliminary results presented in this chapter were published in [Krasnodębski, 2016b].

can be replaced in a certain proportion by resin acids. These resin soaps were popular in resin-producing regions on both sides of the Atlantic Ocean, and, in principle, they exhibited similar properties to the fatty acid ones. However, incorporating resin acids into soaps created a number of terminological challenges. Were these resin-based substances proper soaps or, rather, soap substitutes? Was resin an ingredient on its own merits or, from the consumer's point of view, an undesirable pollutant? Although on the periphery of the Institute's actual studies, both Georges Dupont and Georges Brus actively followed these debates and provided a platform for soap advocates in the pages of the Institute's bulletin.

Starting in the 1890s, an increasing number of soapmakers from Marseille began incorporating a certain proportion of rosin into their soaps in place of fatty acids [Merklen, 1934, p. 212]. Marseille had been a soapmaking centre for hundreds of years, and its famous Marseille Soap 72% extra pure was considered a top-quality product. The figure 72% refers to the proportion of the soap, strictly speaking, in the final product. In practice, in the Marseille soaps, the weight of acids corresponded to 63 per cent, of sodium compounds to 9 per cent, and of water and perfumes to 28 per cent. Soaps became a very important market outlet for the Landes producers, and the consumption of resin in the trade grew continuously until World War I. This development was, however, marked by controversies. A powerful lobby of producers questioned the possibility of incorporating resin acids into the Marseille Soap 72%. Many believed that resin acids lowered the quality of the soap; hence, the brand "Savon de Marseille 72% extra pur" should be limited to the ones using traditional materials, such as palm, coconut and peanut oil. In 1907, a local trade court found a soapmaker guilty of fraud for selling a 72% extra pure soap that had 10 per cent of resinic acids [Portmann, 1932]. The court considered that the resin was a "hidden vice" and that its presence should have been clearly indicated on the packaging. In response, a number of soapmakers invented a range of alternative names to avoid the ruling. Instead of "extra pure" soaps, they sold "72% active matter" ones or "72% oils and saponifiable materials" [N. N., 1924]. Meanwhile, many producers simply ignored the ruling and continued to incorporate some proportion of cheaper resinic acids into their products.

The real conflict began in 1924 when a certain A. Baud, from a departmental laboratory in Marseille, issued a critical report on the use of resin in extra pure soaps. He advocated for banning this practice to safeguard the reputation and standing of Marseille soaps among consumers [Merklen, 1928]. Many local trader unions supported the idea of strict laws concerning soap-related terminology, but there were also defenders of the resin's usage. One study explained that while too much resin acid might make soap sticky, the inclusion of 5 per cent of resin in the Marseille soaps 72% extra pure had no effect and should be allowed [N. N., 1924]. In 1925, a trade court in Lyon exonerated a soapmaker selling resinous soap by explaining that no regulation existed as to what exactly the 72% referred to. This decision angered Marseille soapmakers, who called on the national Fraud Repression Service to intervene. At first, the Service issued a very strict regulation limiting the use of resins, but it was judged illegal by the French State Council since soaps were not part of the Service's jurisdiction (the law allowed to regulate almost exclusively food products) [Dupont 1928, p. 56]. The Service then proposed to the French Parliament to prepare a law regulating the soap market. The first version of the new law was voted on 16 March 1928 by the Chamber of Deputies but was rejected in the French Senate many times in the following years. The bill underwent several revisions and debates in different committees until the early 1930s [N. N., 1932].

The proposition from 1928 did not impose any restriction on the use of resin in soaps in general. However, if resin acids exceeded 5 per cent of the weight of the acid part of the soap (so around 3.4 per cent of the total weight), it could not be sold with the word "pure" in the name [Dupont, 1928, p. 58]. At first, the official position expressed in the Pine Institute Bulletin by its director, Dupont, was favourable. He argued that, in principle, resin soaps could be now safely sold and that the proportion of 5 per cent in the Marseille 72% extra pure soap was a pretty reasonable success. That same month, the chief editor of the local journal *Bois et Résineux* published a much more nuanced article, explaining that it was a pity that Parliament did not want to accept a full equivalency between resin and fatty acids but that the overall direction was good [Bodin, 1928]. And yet, only months later, in June 1928, one local trade organisation criticised the project and claimed that France was the only country in the world limiting the use of resin in soaps. In August, a Marseille soapmaker and defender of resin soaps, François Merklen, raised the tone by explaining that this regulation did not "satisfy anybody, a loud minority aside" and that it was "disastrous" for the resin producers. He notably insisted that these soapmakers who wanted to use resins had already incorporated 5 per cent for many years, and the regulation would only discourage newcomers by stigmatising resins [Merklen, 1928].

The following year, in 1929, one local entrepreneur questioned the purpose of the legislation itself: "Do we have a right, do we even have the power to legislate on such an industrial problem? Are we going to regulate in the Parliament the amount of cocoa to put in chocolate, of sugar in praline, of honey in gingerbread?" For him, the *laissez-faire* principle should reign, and it was up to consumers to choose. He notably believed that the opposition of soapmakers to resin was irrational [Poisson, 1929]. Reunited around *Bois et Résineux*, local *résiniers* and small entrepreneurs fought back against the legislation and rejoiced when the Senate decided to send the project back to the Chamber of Deputies [N. N., 1930].

An intriguing aspect of these discussions is that the participants frequently employed the language of progress to support their respective viewpoints; for example, Merklen wrote that, in the first half of the nineteenth century, Marseille soap had been manufactured exclusively from olive oil. This was the soap that launched the popularity of Marseille products in France [Merklen, 1928]. In the 1850s, sesame grain oil was also envisaged by "clear-sighted manufacturers", but "the 'sacred tradition' ... has risen against the introduction of a new oil. Replacing a proportion of olive oils by grain oils was considered an assault against the soapmaking itself". Those who wanted to use sesame oil had to move the grains to production plants by night in order to evade inspections.

Once ready, the sesame soaps had the same quality as the ones from olive oil. Eventually, the practice became so widespread that by the end of the nineteenth century, using grain oils was no longer seen as fraudulent. Subsequently, the arrival of coconut oils (copra) upset the soapmakers, and a similar situation unfolded once more. According to Merklen, advancements in techniques and chemistry facilitated the incorporation of diverse raw materials, which were both cost-effective and of the same quality, into soaps. This trend also applied to resin soaps. Such developments represented a natural progression in the industry's evolution.

In 1930, pro-resin senator Louis Pasquet, in his ample report on the place of resins in soap, wrote that "with every progress of science, we must strive to overcome the old routine; we have to oppose habit and inertia to innovations ... The precursor is often treated as a deceiver, a fraudster, sometimes even a thief". He added later that "[o]ne cannot dream of subjecting science to regulations ... Limiting, in the industrial products, the proportion of diverse elements entering their composition is equal, in many cases, to the fight against the progress itself" [Pasquet, 1930]. For resin advocates, the entire affair was not a small trade dispute but a part of the great struggle between forces of progress and obscurantism.

Some remained unconvinced. In 1931, a Marseille deputy, Henri Tasso, accused the resin industry of hypocrisy. Why, he asked, did the turpentine producers deny the soap-makers the right to preserve their own trade traditions while simultaneously fight for a regulation to protect turpentine against white spirit [Bouyssou, 1931]? As explained later, Tasso's remarks had a grain of truth in them.

In the early 1930s, the law was finally discarded, a decision celebrated by local producers who hoped it would mean liberalisation. And yet, at the same time, the purview of the Fraud Repression Service was extended, and it often intervened in the soap market, applying a rather strict definition of the 72% extra pure soap, preventing, *de facto*, the use of more than 5 per cent of resin acids. A special commission was established to help the resin industry recover from the crisis and managed to negotiate an increase of this limit up to 8 per cent with the Service in 1936 [Marcard, 1936]. Was this considered a victory for resin producers? On the contrary, in the middle of the 1930s, the crisis ravaging the Landes was at its peak. The President of the Chemical Engineers Union of the South East, René Marcard, called this decision "a platonic invitation". He advocated for a law that would force all soapmakers to incorporate at least 10 per cent of resin in their soaps regardless the marketing name (including 72% extra pure) in order to save the resin industry. Let us reiterate: in the 1920s, the people of Landes were fighting for a law protecting the freedom of choice of soapmakers who wanted to use resin. In the 1930s, they were fighting for a strict regulation imposing its use in all soaps, even against the will of soapmakers.

In the end, no new law was enacted in the interwar period. In the Landes region, resin soaps remained popular, but they got no national traction until World War II, when, due to shortfalls of plant fatty acids, resin acids were officially admitted in soaps on an equal basis [Brus, 1940]. The resin soaps consumption rocketed sky high and soapmakers still consumed around 30 per cent of all French rosin in 1951.[260] However, the resin soaps manufactured during the war were of such poor quality that resin slowly became synonymous with low-quality products. Brus admitted that resin soaps darken rapidly because of oxidation and that the use of hydrogenated rosins would solve the problem [Brus, 1940]. However, these rosins were too expensive to be used in common detergents. After the war, Marseille soapmakers were eager to abandon resins because the consumers associated them with war shortages. By 1970, resin acids almost completely disappeared from the French soap industry.

Standardisation of Rosin Grades

Although the Pine Institute had to acknowledge the existence of an independent technological and industrial tradition beyond its field of expertise in the case of soaps, it was the ultimate authority in France when it came to rosins.

---

[260] Departmental Archives of Gironde, Pine Institute Archives, file "Saint-Symphorien". R. Bentejac, "La gemme, sa production, son utilisation", 1971, p. 31.

As noted, rosin is a solid component of gum resin distillation made of 90–95 per cent of resin acids of a general formula $C_{20}H_{30}O_2$. It always contains some residual turpentine, oxidation products, and many other substances present in wood. In other words, it is not a well-defined material from a chemical standpoint. However, the rosin trade was helped by the fact that for decades, its price depended on its colour and not strictly on the chemical composition. The French even gave a different name to dark rosins (*brais*) and bright ones (*colophane*), but the colours of rosins constituted a continuum with light brown substances standing somewhere in the middle. Colouration does not only depend on the proportion of different acids but on many different factors, such as the presence of organic impurities during the distillation or the rust in the containers.

The uses of the two categories of rosins were not identical. For soaps, especially the soaps from Marseille, the brighter the rosin, the better. However, for road coating or low-value oil manufacture, the colour did not matter, and these industries typically used darker shades. Since the eighteenth century, the industry had prioritised the discolouration of bright rosins over dark ones, as the former was much more expensive. Prior to the discovery of chemical processes for the removal of tannins and other impurities, the method used to brighten the rosins involved prolonged exposure to sunlight in wide, shallow containers (Figure 56).

This primitive method was, nevertheless, not very efficient as the rosins brightened that way tended to darken faster afterwards because of oxidation, not to mention that it required significant costs for the labourers taking care of them.[261]

Because of the impact of their colour on the final price, rosins started being categorised according to their shade. Before World War I, every larger Landes enterprise had its own scale (*gamme* in French) of colour shades used to establish the value of a given batch. These different scales were inconsistent, not to mention that

[261] Departmental Archives of Gironde, Pine Institute Archives, file "Lettres Techniques 1956–1957", letter C. Letter to Centre Français de Renseignements Techniques Industriels, 6 September 1957.

Figure 56: Rosin brightening with sunlight in 1921.

Source: [Brus, 1940, p. 6]

the reference cubes were made of rosin, whose colour could change over time [Lombard, 1946, p. 255]. In 1925, the Pine Institute, together with local producers, established an official scale of rosin grades to remedy this problem.[262] Legitimised by the scientific authority of the Institute, the standard spread throughout the region, becoming a part of every rosin-related transaction. The official scale was made of glass slides of clearly defined thickness and colour. The procedure for measuring the colour was relatively simple. A rosin sample should be of a cube with a side length of 22 mm, placed side by side with a given glass slide on a perfectly white background. A special tool, a colophoscope (see Figure 7), was developed to facilitate the comparison [N. N., 1931]. The full range of glass slides was, however, very expensive as it could cost up to

[262] Departmental Archives of Gironde, Pine Institute Archives, file "Colophane — documents étrangers". "Quelques caractéristiques des colophanes et brais provenant de la gemme de pin maritime", no date.

1,000 francs, so less precise scales made of rosin cubes of different colours were also sold (but had to be replaced every year because of oxidation).

It is notable that the book of a Strasbourg chemist René Lombard, *Produits résineux, gemmes, colophanes & dérivés* (1946) [p. 255], as well as internal documents of the Pine Institute and many articles in the Bulletin [N. N., 1931], presented the Pine Institute as the true innovator whose classification regularised and facilitated the rosin trade all over France.[263] The problem was that the basis for the rosin grades was not originally established in France but in the United States (US). The French system comprised 16 colour grades, ranging from 7A (best quality) to E (lowest quality): 7A, 5A, 3A, 2A, Y, X, WW, WG, N, M, K, I, H, G, F, and E. The American system had 14 grades: X, WW (water white), WG (window glass), N, M, K, I, H, G, F, E, D, C, and B. The two ranges largely overlapped, but the French system produced brighter rosins due to the filtration of gum before distillation. In principle, however, the French scale was a replica of the American one developed by the USDA [Outland, 2004, p. 157] (Table 12).

The first rosin grades had been elaborated in the US already in the 1850s. Many local producers' associations tried to standardise them, and in the first years of the twentieth century, different competing regional scales were used, *i.e.*, the Southern Grading scale or the New York one.[264] This, along with the use of imperfect oxidisable cubes, resulted in half of all rosin samples being incorrectly graded on local markets, according to the USDA's evaluation in the 1910s. As a consequence, from 1912 on, the USDA's naval stores laboratory investigated the matter and tried to unify different

[263] Departmental Archives of Gironde, Pine Institute Archives, file "Colophane — documents étrangers". "Quelques caractéristiques des colophanes et brais provenant de la gemme de pin maritime", no date.

[264] US National Archives (College Park), Records of the Bureau of Agricultural and Industrial Chemistry, entry 69, ARC ID# 3887383, Subject Files Relating to Naval Stores Research, 1903–1935, file "Duplication of work and cooperation", 00530: 052/025/00, box 8. "Memorandum of the leather and paper laboratory projects", 8 December 1916.

Table 12: Rosin grades in the 1920s

French name	Grade	English name	Colour
Brais noirs	B (only in the US)	Common Strained	Black
	C (only in the US)	Strained	Dark brown
	D (only in the US)	Good Strained	
	E	n°2	
	F	Good n°2	
Brais clairs	G	Low n°1	Brown
	H	n°1	
	J	Good n°1	
Colophanes	K	Low pale	Yellow/transparent
	M	Pale	
	N	Extra pale	
	WG	Window glass	
	WW	Water white	
	X and higher	Extra	

systems.[265] In 1914, naval stores inspectors were equipped for the first time with glass slides to monitor the rosin colour within the trade.[266] The laboratory continued to work on different scales in the following years.

When developing grades for rosin, there were various considerations to bear in mind. They had to be numerous enough to incentivise the producers to adopt better cleaning methods, and the

[265] US National Archives (College Park), Records of the Bureau of Agricultural and Industrial Chemistry, entry 69, ARC ID# 3887383, Subject Files Relating to Naval Stores Research, 1903–1935, file "Duplication of work and cooperation", 00530: 052/025/00, box 8. "Leather and Paper Laboratory 1913".

[266] US National Archives (College Park), Records of the Bureau of Agricultural and Industrial Chemistry, entry 69, ARC ID# 3887383, Subject Files Relating to Naval Stores Research, 1903–1935, file "Duplication of work and cooperation", 00530: 052/025/00, box 8. "Memorandum of the leather and paper laboratory projects", 8 December 1916.

same category should not include rosins on too wide a spectrum. At the same time, too many grades would have required the use of specialised equipment for the colour identification. No sophisticated colophoscopes were sold on the American market, and the Americans relied on much simpler and less precise tools.

National standardisation was not always welcome though. During the debates on the Naval Stores Act in the US Senate, one senator from Georgia argued that his state already had a classification system and did not need a new federal one. He believed that since the Savannah market was the biggest in the US, other producers could simply use the Savannah grading scale instead of the one approved in Washington [N. N., 1919, p. 53]. At the same time, the proponents of the federal system extolled the unalterable glass slides that were much more practical than the rosin cubes distributed at the Savannah market. These slides were, in fact, manufactured by the British company Lovibond-Tintometer, and were praised for their colour quality and precision.

Another argument in favour of a binding federal regulation on rosins was that it would permit the generation of comparable trade data at the national level, a task impossible with many competing grading systems. Of course, this would also require giving control to the federal authorities. The state governments vehemently opposed this encroachment, but the majority of producers were in favour of the standardisation to facilitate and harmonise trade. Fletcher Veitch, representing the US Bureau of Chemistry during the hearings, added that the new grades were not trying to revolutionise the system but to build upon and improve it, and that they were, in fact, similar to the ones used in Savannah [N. N., 1919, p. 18]. Eventually, the Senate was convinced and approved the Naval Stores Act in 1923, making it mandatory to use the federal grades on American territory [Veitch, 1924].

In the same fashion as the USDA constructed its scale based on the Savannah system, the French Pine Institute created its own system, expanding on the American one. Its true contribution was the definition of the brightest grades that had been simply categorised under the letter "X" in the US. In its early publications, the Institute called the lower grades the "American ones" and the upper grades "the French ones", but this differentiation was soon forgotten, and subsequently, many seemed to believe (in France at least) that the Pine Institute was the inventor of the entire system. While the French were not eager to highlight the fact that their system was based on the one developed by their greatest competitor, the Americans noted this parentage link with great pride. An internal USDA report from 1930 explained that:

> Until recently, the naval stores trade of France has had no definite and permanent standards of rosin. Through cooperation with Dr. Dupont and other French leaders, the French trade has now adopted as official for French rosin the colour values and nomenclature of the United States standards from B to WW inclusive. Steps are being taken to bring into agreement the paler shades, *i.e.*, the American X and French Y grades.[267]

Both the French and the Americans used the glass slides manufactured by Lovibond-Tintometer. In France, while the cooperation appears to have been fruitful in the 1930s, and the Institute often provided coordinates of the society to producers or acted as an intermediary for acquisitions, this changed after World War II. From the 1940s until the 1960s, national and foreign companies regularly consulted the Institute about the French scale. But the answer was always the same: "We do not have the rosin grading scale in stock at the moment."[268] The Institute explained that Lovibond-Tintometer imposed prohibitive prices and huge delays, and that negotiations were underway to solve these difficulties. As

---

[267] US National Archives (College Park), Records of the Bureau of Agricultural and Industrial Chemistry, entry 69, ARC ID# 3887383, Subject Files Relating to Naval Stores Research, 1903–1935, 00530: 052/024/00, box 4, file "Memoranda 1927 — June 30 1935". "Naval Stores Investigations".

[268] For example: Departmental Archives of Gironde, Pine Institute Archives, file "Colophane — normes étrangères". Letter from Sociedade Continental de Productos Resinosos, 30 September 1954.

a temporary solution, the Pine Institute advised its interlocutors to order grading cubes made of rosin from the U.C.R.[269] Indeed, Lovibond-Tintometer wanted in the early 1950s, no less than 50,000 francs for a single scale, accounting for inflation, it was twice as much as before the war. Why was it so expensive? There is no clear answer, but there is one possible hypothesis. In 1936, the Americans adopted a new standard made of 12 grades: X, WW, WG, N, M, K, I, H, G, F, E, and D. At first glance, they seemed similar to the old ones, but in reality, the entire system was thoroughly transformed. The space between the upper and lighter grades was reduced, meaning that the X and WW were considerably closer to each other than E and D. This was made to meet the growing requirements of increasingly picky industries looking for clear grades, as well as the relative loss of importance of lower grades [Scofield, 1941, p. 20]. In other words, the American grading system was subtly but deeply reconstructed. It would explain why the Pine Institute ordered the American scale in 1941 to study it.[270] Everything suggests that the French grading slides became a rarity after the war, and that their manufacture was less financially rewarding for Lovibond-Tintometer, which was already supplying the Americans with their new grades.

In summary, the French continued to operate *de facto* with the American system from 1923, while the Americans themselves moved away from it. The French did not revise their own grades until the 1950s when the French scale was expanded by two new brighter grades (beyond 7A): "cristal" and "super cristal".[271] A few years later, in 1968, the Americans expanded their own revised grades with three new levels: XA, XB and XC [Gardner and Sward, 1972, p. 82].

[269] Departmental Archives of Gironde, Pine Institute Archives, file "Correspondance 1949, B". Letter from Savonnerie de la Bouche, 24 May 1949.
[270] Departmental Archives of Gironde, Pine Institute Archives, file "Correspondance 1941, E". Letter from the US consul, 14 February 1941.
[271] Departmental Archives of Gironde, Pine Institute Archives, file "Lettres Techniques 1952–1953, S". Letter to Institut du Pin, 18 July 1953.

Interestingly, both scales were still mentioned on the website of Lovibond as of 2023.[272] The company sells rosin comparators adaptable for two scales. The American one is made of 15 grades (XC, XB, XA, X, WW, WG, N, M, K, I, H, G, F, E, D), whereas the French has 10 (super cristal, cristal, 7A, 6A, 5A, 4A, 3A, 2A, Y, X). It means that the French scale, strictly speaking, does not include grades below X. However, while the grades below X were simply from the American scale until 1936, there was no full compatibility anymore after that year. In other words, a curious and somewhat incoherent translation of standards and technological solutions one hundred years ago can still potentially confuse consuming industries until this day.

2. Defining Turpentine Spirit in France

The case of turpentine is perhaps the most emblematic one when it comes to the Institute's activity in standardisation. It illustrates the influence of scientific institutions on national and international trade, but it also shows a complex relationship between what constitutes scientific definitions and trade norms. In order to understand the issues at stake in the twentieth century, this section first examines the etymology of the word turpentine and its earliest uses.

Origins of Unclear Terminology

The name "turpentine" (*térébenthine* in French) originated from the tree *Pistacia terebinthus*, known under the name "terebinth" (or turpentine tree) until today. It is a deciduous tree widespread on the eastern coasts of the Mediterranean Sea and in the Middle East. The name *terebinthus* appeared as early as in the writings of Pliny the Elder and Dioscorides. The latter wrote that terebinth resin was

[272] Lovibond's website, with information on rosin scales (https://www.lovibond.com/en/PC/Colour-Scales/Rosin-French-Bordeaux-Colour, last accessed 18 February 2023.)

better than any other for treating cough and tuberculosis [Dioscorides, 2005, p. 630]. Its applications in antiquity were, above all, medical. The famous Chio turpentine from terebinths growing on a small island in the Aegean Sea was sought after in all the major trade centres of the Roman Empire. And yet, already in antiquity, the word turpentine was occasionally used to describe other resins with similar properties [Aufan and Thierry, 1990, p. 8].

In the first French manual of chemistry by Nicolas Lémery from 1675, the author explained that there were "two types of trees from which turpentine comes ... terebinths and Larix [larches] ... Turpentine is, strictly speaking, a liquid resin of balsamic consistency; the one coming from the isle of Scio [or Chios] is considered the best" [Lémery, 1675, p. 485]. He then described how to extract turpentine spirit from turpentine by distillation. Seventy-five years later, the *Encyclopaedia* by Diderot and d'Alembert offered a comprehensive summary of what the turpentine was according to the consensus of the period. It explained that turpentine was a special type of resin coming from certain trees. It enumerated five types of turpentine. The most expensive ones were Chio and Persian turpentines secreted by *Pistacia terebinthus*. According to eighteenth-century scholars, terebinths remained the source of the highest quality turpentines. Other types of turpentine included Venice turpentine, secreted by a species of larches in the Apennines; Strasbourg turpentine, from the silver firs of Alsace; and, finally, common turpentine, of the lowest quality, produced from pines.

In the Landes near Bordeaux, a different vocabulary emerged in the fourteenth century. On the markets of Bordeaux and Dax, words such as "terbentina", "tormentine", and later "tourmentine" and "turmentine" were frequently used [Aufan and Thierry, 1990, p. 77]. What matters is that these terms did not refer to raw resin extracted from pine trees but to a volatile liquid appearing in the pine resins in a heat treatment process. One of the ways to obtain this *térébenthine* involved pouring raw resin into huge, shallow, open boxes and then placing them under sunlight. Under the influence of solar heat, a clear division became noticeable between the solid part on the bottom, *galipot* (made of rosin and polluting

agents), and the liquid "sun turpentine" (*térébenthine de soleil*) on the surface [Aufan and Thierry, 1990, p. 14]. This *térébenthine* quickly found its uses as a solvent in varnishes, paints, and sealing waxes. The sun turpentine was a very clear resin with a huge proportion of terpenes, standing somewhat between turpentine spirit (understood as a resin distillate) and a raw pine gum gathered at the tree. Its remarkable properties encouraged local inhabitants to discover new production processes, and they quickly learned to heat pine resin in cauldrons to obtain syrup that was richer in "essence" than sun *térébenthine*. In 1709, the Bordeaux chamber of commerce defined *esprit de térébenthine* as the product obtained from the distillation of pine gum resin in an alembic. By 1749, it was possible to buy a set of tools necessary for the proper distillation of turpentine oil (*huile de térébenthine*) in Aquitaine. The two-step resin treatment method emerged in the region organically. Cleaning heated resin (*térébenthinage*) and the distillation, strictly speaking, were not fully independent steps in the seventeenth and the eighteenth centuries.

This short introduction was to identify the roots of two major further terminological debates. Firstly, the original meaning of the word "turpentine" referred for hundreds of years to the resin of *Pistacia terebinthus*. Other resins were sometimes called turpentine by analogy because of similar properties (both in medicinal uses as well as in varnish-making). It was due to the rapid expansion of the naval stores industry that the term turpentine became associated with pines throughout the nineteenth century and largely replaced older meanings. Secondly, according to consensus in the eighteenth century, the term turpentine referred exclusively to crude resin and not the distillation product. It was exceptionally in Aquitaine where *térébenthine* acquired a distinct meaning of a purified gum that could be transformed into *essence de térébenthine*. Then came a terminological divergence that would haunt generations of chemists in the French and English-speaking worlds for decades. In France, the etymology was clear; *essence de térébenthine* meant turpentine spirit. There is a substance called *térébenthine*, obtained by *térébenthinage*, which after distillation becomes *essence de térébenthine*. In the US, however, there was no *térébenthinage*. As a consequence, the terminology became much laxer. American chemists and regulators used the name "turpentine spirit" in the early twentieth century, but the naval stores businesses regularly referred to it simply as turpentine. In practice, over the years, the terms "turpentine spirit" and "turpentine" became interchangeable even in the official use. Today, bottles of turpentine can be easily purchased in the US, but *térébenthine* is not sold in France, as the French consistently use the name "*essence de térébenthine*" for the gum's distillate.

Turpentine Standardisation in France in the Interwar Period

As previously explained, one of the most important contributions of Maurice Vèzes' laboratory was a deepened analysis of the composition of *essence de térébenthine*. The laboratory's chemists succeeded in developing simple methods and procedures to differentiate turpentine spirits of different origins and detect adulterants, which enabled much more precise quality control. The nature of frauds had evolved over the ages, but around 1900, the most common ones involved adding water or white spirit (low-quality, oil-based solvent) to *essence de térébenthine* or distilling pine resin in an improper way, leading to an essence with too much resinic acid. Nonetheless, for countering frauds, quality control was not sufficient. Establishment of clearly defined thresholds that a proper turpentine spirit should satisfy was necessary. In other words, the sector was in need of standardisation.

In 1909, during a Congress of Fraud Repression Service in Paris, Vèzes formulated one of the first rudimentary scientific definitions of turpentine spirit applicable to commerce. The definition required distilling resin with water or steam without overheating, and established a few general parameters to identify adulteration with white spirit [N. N., 1924b]. At the time, it was known that pine resins were made of poorly known resin acids and of terpenes. The main terpenes in turpentine spirit were α- and β-pinenes in the case of the most popular pine species (maritime pines in Aquitaine, longleaf and slash pines in the US). However, other terpenes, such as limonene or sesquiterpenes, were also identified, and the proportions of α- and β-pinene varied significantly between different pines. What is crucial is that the distillation of pine resins never gave pure terpenes. A certain quantity of resin acids and other impurities was always present in turpentine spirit, depending on the quality of distillation and of the purification of the resin before the distillation. In this sense, every attempt to define *essence de térébenthine* involved establishing an arbitrary boundary between a properly distilled substance and an "impure" one.

Ironically, the definition from 1909 was swiftly deemed impractical by the laboratory scientists due to two major flaws. Firstly, it failed to consider that turpentine was still being manufactured in many workshops using primitive direct distillation apparatus (*à cru*). As a result, it was distilled unevenly. Secondly, it proscribed the use of new fractionation methods that did not involve steam or water. The definition, thus, was not well adapted to the requirements of both the least and most innovative production systems (N. N., 1924b).

While the definition proposed by Vèzes was not legally binding, the French Fraud Repression Service was already asking Vèzes' laboratory to detect fraud in resinous products in 1908. In effect, the laboratory became the sole authority in France to define *essence de térébenthine*. The laboratory's chemists, strongly linked to the local industry, acquired the power to impose their own ideas about this product at the national level. The specialists at the laboratory in Bordeaux knew traditional practices from the local markets, and the main goal of the laboratory was to provide analytical services and support to local merchants and producers, thus adhering to trade definitions that predated scientific analysis. Although the local communities developed certain trade and business practices that could be legitimately sanctioned without challenge, problems appeared with the imported products of different origin [Barraud, 1930]. What were they to do with solvents sold as *essence de térébenthine* but manufactured with methods that were not used in the region and from different pines (not to mention even from different coniferous trees)? As noted in Chapter 5, it was possible to obtain solvents similar in properties to turpentine spirit, not by distilling resin extracted from trees but by distilling resins directly from the pine wood itself by steam distillation of light wood. The French argued that all these substances were simply different products and, scientifically speaking, they were not *essence de térébenthine* at all.

During the French National Congress of Industrial Chemistry in 1924, Dupont, after having worked with a special governmental commission on the standardisation of turpentine, proposed a new set of detailed characteristics to describe resinous products in a much more precise way [Dupont, 1924]. He defined turpentine spirit as a "mix of terpenes, having only a small proportion of sesquiterpenes and oxidation products. It is a colourless liquid, sometimes slightly greenish or yellowish, very fluid, of a characteristic smell, and its density is never smaller than 0.860 at the temperature of 15 degrees". He then added that it started boiling at 152 °C and "gives" at least 90 per cent of its weight below 170 °C. Dupont further suggested that the proportion of small amounts of rosin and resin oil should not exceed 2.5 per cent of the weight. These simple thresholds, consistent with the established practice, rapidly gained widespread approval in the region. However, they had to wait until 1931, the middle of the most severe crisis to affect the resin industry in Aquitaine, to become not only scientifically but also legally binding. On 30 December 1931, a new law on "the repression of frauds in the commerce of *essence de térébenthine* and resinous products" was issued, followed by a decree and a ministerial circular defining precisely what was hiding behind the term *essence de térébenthine* [N. N., 1932b]. Recall that such legislation was necessary because the Fraud Repression Service was not, in principle, allowed to regulate non-food products.

The road to legislation regarding turpentine was long, controversial and complex, much like the journey of Marseille soaps. It started in 1924 when a trade court in Poitiers acquitted a merchant who sold a "Turpentine, Swedish essence, purity guaranteed", even though its composition and properties were vastly different from the one produced in the Landes (it was most likely a poorly distilled sulfate turpentine from paper mills). The Landes producers' association criticised the decision and called for abolishing the use of the name "turpentine" on products imported from Sweden, claiming they were not "real" turpentines [Rouin, 1924]. Dupont himself considered these substances, which he called pine spirits (*essence de pin*), very dangerous to human health, in a comment to Rouin [1924].

Foreign turpentine substitutes aside, the local producers also fought against white spirit, underlining that it should not be used to dilute turpentine, and that it should not be marketed as mineral turpentine as some sellers tried to. A certain J. Capus wrote in *Bois et Résineux* in 1931:

> One should not say, as I have heard sometimes, "the substitution of turpentine by white spirit is one of the consequences of the progress of science, before which we have to bow" ... the sole fact that some new process is of scientific origin does not mean at all that it constitutes a technical progress ... The slavish acceptance of all new processes would lead us, under the pretext of progress, to a civilisation of quantity, of junk, which would lead to all sorts of decadence [Capus, 1931].

This epistemological musings in the pages of a forestry journal reflects the zeitgeist reigning in the region. In 1927, André Bodin, the editor-in-chief of *Bois et Résineux*, warned against excessive enthusiasm for scientific discoveries: some "confuse, in the same type of admiration, the effort of the scientist, which is sublime, and the practical consequences of his discovery, which are sometimes unexpected ... The science resembles the chariots of these Hindu deities that crush the believers on their passage" [Bodin, 1927]. How could this mistrust towards innovation and the firm belief in the necessity of progress to save the resin industry from the crisis be reconciled? Again, Capus provided some answers: "We can say today that the legislation against frauds, as well as an appropriate agency that would take care of analyses and inspections for ensuring its application, is necessary ... in order to help the civilisation to protect itself against false progress and false economies." [Capus, 1931] Capus advocated in favour of the new law for turpentine protection and saw the Pine Institute as its future guardian. For local resin producers, the regulation of the amount of resin in soap was a false misguided progress, which had to be resisted, while the strict regulation of the origin of turpentine was the defence of the "true" science in service of humanity.

Whatever the ideological justifications were, the new regulations (law, decree and circular) simply reproduced the results and definitions that had been elaborated by the Pine Institute throughout the 1920s. The law abolished:

> [U]sing the name *essence de térébenthine* and every other expression with the word *térébenthine*, as well as combinations, derivatives, or imitations of this word to designate any product not coming presently, directly and exclusively, from distillation, at a temperature below 180 degrees, of oleoresin juices obtained by tapping a variety of different living pines, which are cultivated according to loyal and constant custom to produce *essence de térébenthine* [N. N., 1932b].

The subsequent decree added: "*Essence de térébenthine* is a colourless liquid whose density is never less than 0.860 at the temperature of 15 °C; it is made of terpenes; it begins boiling ... above 152 °C and during the distillation has to lose at least 90% of its weight before reaching the temperature of 170 °C."

There are numerous interesting facets here. The law on resinous products proscribed in practice the use of the term "turpentine" and any of its variants for products that did not contain *essence de térébenthine* as defined by the law. In other words, it was not possible to sell, for example, merchandise called "turpentine oil" unless it satisfied the same criteria as *essence de térébenthine*. What was even more troubling is that the term *essence de térébenthine* was divorced from its etymological origin since, according to the law, it should be distilled from oleoresin, not from *térébenthine*. As such, the term *térébenthine* became somewhat an empty shell, legally meaningless.

Not only that, this *essence de térébenthine* was also supposed to come exclusively from pine resin, other species having been strictly excluded. It was only an administrative circular published in 1932 that permitted the exceptional use of the term "*térébenthine*" in commerce to describe some products that had been traditionally called this way, such as, Venice turpentine from larches and Strasbourg turpentine from firs, as well as in the meaning given to it by Landes forest workers (purified resin). As a consequence, even though the word "turpentine" came from *Pistacia terebinthus*, the source of Chio turpentine, Chio turpentine ended up on the list of exceptional uses in France in the 1930s. However, if a merchant sold *Pistacia terebinthus* resin sourced from a different region, such as Palestine, where the tree was abundant, they were not allowed to market it as "Palestine turpentine" on the French market.

Another interesting aspect was that *essence de térébenthine* could only be produced from trees that were tapped "according to loyal and constant custom". This reasoning was obviously circular and had clearly protectionist aspirations. In France, only maritime pine trees satisfied this criterion.

However, the most crucial measure adopted by the law concerned the terminology used to describe other substances similar to turpentine spirit but is ineligible to be labelled as *térébenthine*. One such term was *essence de pin* (pine spirit), which, according to the law, was distilled from pine wood (not resin) or obtained through processes other than traditional pine resin distillation, such as those used in paper mills. Its quality thresholds were also lower than for *essence de térébenthine*: its boiling point should be at 150 °C, and at least 85 per cent of its weight should be distilled below 190 °C. The second category was *essence de bois* (wood spirit), which could originate from any coniferous wood no matter the extraction method and should give 80 per cent of its weight between 80 °C and 200 °C.

This resulted in a double classification system that depended on both the origin of the substance and its properties. This means that even if a spirit obtained from pine wood distillation satisfied the threshold fixed for the normal *essence de térébenthine*, it could not use this label. In other words, the names "*essence de pin*" and "*essence de bois*" hid a whole range of substances of different origins with very different properties. A French customer, when faced with a choice between *essence de térébenthine* and *essence de pin*, was inclined to choose the first, since the law guaranteed its quality, but, in theory, the latter might have had exactly the same properties. When a foreign high-quality product of wood origin was sold on the French market, it had to use the label that made it indistinguishable from low-quality products.

This regulation might be seen as very conservative. By using the word "directly" as referring to distillation from gum, it effectively abolished the usage of the name "turpentine" to any wood turpentines from the US. However, this conservatism was attenuated by the use of the word "presently" (*actuellement* in French). In fact, it opened up the possibility of using methods other than distillation to obtain turpentine. It is probable that the Pine Institute had considered already in the 1930s the possibility of direct gum fractionation that it developed in the 1940s. This fractionation did not involve distillation; therefore, if it weren't for this single word, the "fractionation turpentine" would not be *essence de térébenthine* according to the law.[273] It shows how deeply the Pine Institute was involved in crafting every single word of the new standard.

The Pine Institute had unquestionable authority over the matter in France. Its definitions were considered objective and unbiased, because they were based on meticulous scientific research. However, on the other side of the Atlantic Ocean, chemists working for the naval stores industry built a nomenclature considerably different from the one elaborated by their French counterparts, and also claimed it was rooted in science.

Turpentine Definitions in the United States and Other Countries

In the US, the first standards for turpentine were issued in 1915 by the American Society for Testing Materials (ASTM). They were later revised in 1924, 1926 and 1934. While ASTM standards were voluntary, a different compulsory regime was established by the Naval Stores Act of 1923, guaranteeing minimum quality thresholds and proper labels on products [Veitch, 1924]. Both regimes were based

[273] The circular completing the law explicitly explained this was the intention behind using the term "actuellement" in the definition of the turpentine [N. N., 1932b].

on the work undertaken in the naval stores laboratory of the Bureau of Chemistry in the United States Department of Agriculture (USDA).

The American standards made a distinction between four types of turpentine. The first category, gum spirits of turpentine, corresponded to the French *essence de térébenthine*. According to the Naval Stores Act, this substance came from the distillation of resins extracted from living pines. Whereas in France, the most popular pine variety was maritime pine, in the US, longleaf pine and slash pine were the dominant species. Their resin gave a product with a very similar composition to the French *essence de térébenthine* and with almost exactly the same properties. Interestingly, gum spirits of turpentine was a traditional name to describe products distilled from resin, contrary to those distilled from wood. The word "gum" (*gemme* in French) was used in this context as a synonym for resin. However, pine resins are not gums, as explained in Chapter 2, according to contemporary conceptualisations [Langenheim, 2003, p. 24]. In this sense, the expression "gum spirits" is quite anachronistic today. It demonstrates once again how convoluted and complex the terminology became.

Three other types of turpentine described by the American industry are significantly different from those used in France. The second type was steam-distilled wood turpentine, meaning "wood turpentine distilled with steam from resin within or extracted from the wood" [Veitch, 1924]. This kind of turpentine used pine wood, particularly old resinous pine stumps and light wood, as a primary resource. In France, the corresponding product was called *essence de pin* (pine spirit). The third category was "destructively distilled wood turpentine", which meant turpentines obtained from wood carbonisation and similar processes. In France, its properties would qualify it as *essence de bois* (wood spirit). The final category was sulfate wood turpentine, a pulp-manufacturing byproduct from the paper industry. In France, it also qualified as *essence de pin* because of its origin. Interestingly, an analysis of ASTM standards shows that the properties of all these products were very close to each other (Table 13). Moreover, in the vast majority of practical applications, all these turpentines were interchangeable.

In the US, the method of extraction was linked to the label, meaning that wood turpentines could not be sold as gum spirits of

Table 13: ASTM Standard D13–1934 on turpentine (extract)[274]

	Gum spirits of turpentine	Wood turpentine: Steam-distilled	Wood turpentine: Sulfate	Wood turpentine: Destructively distilled
Density 15.5 °C (Min–Max)	0.860–0.875			0.860–0.866
Refractive index at 20 °C, D line	1.465–1.478			1.463–1.483
Boiling point (760.00 mm Hg)	150 °C–160 °C			150 °C–157 °C
Distilled below 170 °C	90%			60%

turpentine, very much like in France. However, unlike in France, the requirements for both were very close. They were both "turpentines" having to satisfy the same constraints, meaning that consumers buying turpentines in the US did not need to consider their origin, unless for very specific industrial uses, because their quality was guaranteed by law. The US government's primary interest was ensuring high-quality standards within the country rather than pursuing protectionist policies. Interestingly, the French were more liberal in one aspect: the French *essences de bois* could be produced from all coniferous trees. In the US, pines were the only accepted source of turpentine.

At the beginning of the twentieth century, France and the US were the largest turpentine producers, controlling about 90 per cent of the global naval stores market. Comparing standards adopted by some of the largest consumers of turpentine, such as Germany and the United Kingdom (UK), reveals interesting differences. In Germany, the regulation was somewhat similar to that in France, although it was much simpler. The basic distinction was between *Balsamterpentinöl*, corresponding to the French *essence de térébenthine*, and *Holzterpentinöl*, which can be translated as wood turpentine, similar to the French *essence de pin* or *essence de bois*. Quality thresholds for gum spirits of turpentine were slightly higher

[274] Departmental Archives of Gironde, Pine Institute Archives, box "Normalisation ISO", file "Norme France". "A.S.T.M. Specifications Standard de l'Essence de Térébenthine".

Table 14: Comparison of some properties in gum spirits of turpentine standards. All values are updated for 1949[275]

	France	United States	United Kingdom		Germany
Initial legal basis	Law (1931), Decree, Circular (1932)	Naval Stores Act (1923), ASTM D 13–34 (1934)	Type I B.S. 244 (1926)	Type II B.S. 290 (1929)	R.A.L. Blatt 848 (1927)
Density	$d\,\frac{4}{15}$: 0.860 (min)	$d\,\frac{15.5}{15.5}$: 0.860–0.875	$d\,\frac{15.5}{15.5}$: 0.862–0.872	$d\,\frac{15.5}{15.5}$: 0.859–0.875	$\left(d\,\frac{20}{4}\right)$: 0.859–0.875
Boiling point	Minimum 152 °C	150–160 °C	Maximum 1% distilled below 150 °C	Maximum 1% distilled below 150 °C	152–155 °C
Distilled below temperature	Minimum 90% below 170 °C	Minimum 90% below 170 °C	Minimum 95% below 170 °C	Minimum 70% below 170 °C; min. 90% below 180 °C	Minimum 90% below 170 °C

than for the wood-based one in Germany. In the UK, there was a distinction between Type 1 turpentine and Type 2 turpentine. Type 1 corresponded to the French *essence de térébenthine* but was of very high quality (density between 0.862 and 0.872, and at least 95 per cent of the weight should distil before reaching 170 °C). Type 2 included all the gum spirits of turpentine of lower quality, as well as wood turpentine, regardless of quality (Table 14).[276]

[275] All the values are for year 1949; they may have been revised multiple times after the first standardisation attempt. Adapted from Departmental Archives of Gironde, Pine Institute Archives, box "Normalisation ISO".

[276] Departmental Archives of Gironde, Pine Institute Archives, box "Normalisation ISO", file "Norme France". G. Brus and P. Legendre, "Rapport sur la normalisation de l'essence de térébenthine", 20 May 1949, p. 3.

The key point is that all these nations abandoned the *Encyclopédie*'s vocabulary and opted for industrial definitions that limited turpentine to pine products that satisfied very narrow specifications.

3. International Standardisation of Turpentine (1950–1976)

Upon initial inspection, the specifications concerning turpentine spirit adopted by different countries were relatively close. This was not unexpected since both Germany and the UK imported raw materials from the US and France. At least the differences did not seem insurmountable. And yet, despite all the similarities between national standards, the task of international standardisation committees proved to be more challenging than initially anticipated.

ISO Meeting in Zurich and the French Expertise

The International Federation of the National Standardizing Associations (ISA) planned to standardise turpentine to facilitate international exchange as early as the 1930s. It asked the French Standardization Association AFNOR, to prepare a new international standard on turpentine in 1936.[277] The ISA suspended its activities because of World War II, but AFNOR was aware of the fact that after the end of hostilities, the question of turpentine standardisation would return. It invited Brus, already known as one of the outstanding experts in the field of resin chemistry, to participate in standardisation work.[278] After the war, the successor of the ISA, the International Organization for Standardization

[277] Departmental Archives of Gironde, Pine Institute Archives, box "Normalisation ISO", file "Norme France". Georges Brus and Pierre Legendre, "Rapport sur la normalisation de l'essence de térébenthine", 20 May 1949, p. 3.

[278] Departmental Archives of Gironde, Pine Institute Archives, box "Normalisation ISO", file "Normalisation Essence de Térébenthine correspondance". Letter from Henri Rabaté to Georges Brus, 26 October 1944.

(ISO), renewed its interest in turpentine and Brus voluntarily agreed to participate in the international negotiations. He submitted a proposal for an ISO standard on turpentine, and his suggestions were entirely incorporated by AFNOR into the new French national standard T33-001 published in 1950. This updated the French definition of *essence de térébenthine* and provided a detailed description of the analytical methods necessary for controlling its properties.[279] The same year, AFNOR, representing France in ISO, invited Brus to an ISO meeting in Zurich to present the new standard to the delegates from other countries. The work was done inside the ISO Technical Committee 35, specialising in paints and varnishes.

The meeting in Zurich was very fruitful for the French delegation. The minutes from the meeting clearly indicate that the international participants considered Brus a world expert in the field of resin chemistry.[280] The committee worked using an exhaustive report comparing turpentine standards all over the world prepared by Brus and the Pine Institute's technical director, Pierre Legendre.[281] The French explained that "this report is limited to the gum spirit of turpentine, which is the only one entitled to be named *essence de térébenthine* in France". Brus emphasised that there could be no confusion between genuine spirits of turpentine and wood spirits, and, under his influence, the committee agreed that turpentine spirit should have a separate standard of its own to protect it from abuses and other similar substances.

The French report acknowledged that some turpentines contained terpenes other than pinenes (*e.g.*, limonene in *Pinus pinea*) and that some did not contain terpenes at all (*Pinus sabiniana*

[279] Departmental Archives of Gironde, Pine Institute Archives, box "Normalisation ISO", file "Norme France". "Turpentine spirit standard proposition".

[280] Departmental Archives of Gironde, Pine Institute Archives, box "ISO", file "Zurich". Draft Report of the Meetings of ISO/TC 35 held at Zurich, Switzerland on 26–29 June 1950.

[281] Departmental Archives of Gironde, Pine Institute Archives, box "Normalisation ISO", file "Norme France". Georges Brus and Pierre Legendre, "Rapport sur la normalisation de l'essence de térébenthine", 20 May 1949.

Douglas and *Pinus Jeffreyi*). These and other similar cases were deemed "out of the scope of the definition of typical commercial spirits" according to the French delegates, and, as a consequence, could not be governed by the standard in question.[282] In other words, two independent definitions of turpentine were conceivable: a broader one, based on scientific consensus, including substances containing chemical compounds other than terpenes, and a narrower one, following trade and industrial requirements. No such distinction was made in the French national regulations. This also shows a continuous tension between what constitutes a proper scientific definition and what constitutes a mere trade standard.

One could presume that the French delegation to the ISO, after having recognised the arbitrariness of the proposed categories in its own report, would have been willing to accept more flexible trade definitions of turpentine. On the contrary, only a few paragraphs later, the French report proposed to abolish the use of the term "*essence de térébenthine de bois*" (wood spirits of turpentine) altogether. It was judged to be incorrect because, according to the French experts, *essence de térébenthine* came only and exclusively from the distillation of pine resin, not wood. For the same reason, the French delegation affirmed that the English term "gum spirits of turpentine" and the German's "*Balsamterpentinöl*" constituted pleonasms (just as how the French "*essence de térébenthine de gemme*" would).[283] Brus explained that turpentine was always of a "gum" origin and that using a double qualification might suggest there were other kinds of turpentine. In the conclusions of the report, the French appealed to the American authorities to give up ambiguous terms such as "steam-distilled wood turpentine", "sulfate wood turpentine", and "destructively distilled wood turpentine", and to replace them with the French-inspired terminology of "steam-distilled pine

[282] Departmental Archives of Gironde, Pine Institute Archives, box "Normalisation ISO", file "Norme France". Georges Brus and Pierre Legendre, "Rapport sur la normalisation de l'essence de térébenthine", 20 May 1949, p. 5.

[283] Departmental Archives of Gironde, Pine Institute Archives, box "ISO". Document ISO/TC 35 (Franc — 2) 27, 10.

spirit", "sulfate pine spirit", and "destructively distilled pine wood spirit", respectively. The French delegation also criticised the British regulation for making a distinction between Type 1 and Type 2 turpentines as illogical and, simultaneously, too restrictive and too permissive. It was too permissive because Type 2 also included wood spirits and too restrictive because Type 1 characteristics were applicable only to the purest and best distilled gum turpentines.

The Hague Meeting and a Setback for the French Project

The scope and audacity of the propositions of the French delegation may seem surprising, considering that it was the US that controlled the majority of the production of the naval stores in the world. Indeed, the report was not well-received by everyone. At a subsequent meeting at The Hague in 1953, the ISO Technical Committee 35 decided to revisit the question of turpentine in spite of French protests: "Mr. Dionisi (France) asked if it was necessary to re-address the resolutions of the first meeting. The chairman observed that many new members were present now, and it was decided that the resolutions of the Zurich meeting shall not be considered final."[284] Moreover, contrary to French objections, a new subcommittee was formed to deal with the standardisation of wood turpentine. Even worse, many delegates suggested the possibility of merging the two standards into one, dealing with all turpentines regardless of origin. The possibility of establishing one common standard had already been conceived in 1951.[285] Brus had explained then to AFNOR that such a regulation would be unacceptable because it might "provoke confusion among customers" and "would require developing a new nomenclature for French products, similar to the one in the USA where the word 'turpentine' had much wider

[284] Departmental Archives of Gironde, Pine Institute Archives, box "ISO", file "Haye". Report of the Meetings of ISO/TC 35 held at The Hague, Netherlands on 11–15 May 1953.

[285] Departmental Archives of Gironde, Pine Institute Archives, box "ISO", file "Zurich". Letter from Georges Brus to AFNOR, 30 June 1951.

applications", a solution that was completely unacceptable because it would undermine the "objective" scientific definitions of the Pine Institute.

The issue with having two standards for turpentine extended beyond just naming conventions; for example, the Netherlands and Belgium opted for lowering the required lower limit of density of gum spirits of turpentine from 0.860 (the value proposed by the French) to 0.855 ($d\,\frac{20}{4}$).[286] Brus rejected these demands, claiming that density was a prime factor in differentiating *essence de térébenthine* from *essence de bois* (wood turpentine). The Netherlands replied that the turpentine they had imported from Indonesia had a density of 0.857, and Belgium noted that low density was common for products imported from the Soviet Union. France disputed Belgium's claims and responded that the Soviet Union required a density of 0.860 in its own national standards. France also questioned the authenticity of the gum spirits from Indonesia, doubting whether they were truly obtained from the distillation of pine resin. The Netherlands quickly responded that their turpentine came from *Pinus merkusii* from Sumatra, the density of which was simply lower than that of European and American pines.[287] The Dutch explained that in many parts of the world, "authentic turpentine spirits" had densities between 0.855 and 0.856.

This shows one of the major limitations in the expertise of the French scientists. Their knowledge was built upon samples coming from the pine forests close to Bordeaux; the Pine Institute was mostly unfamiliar with more exotic turpentines, which were rarely sold on the European markets. On the other hand, the question arose as to what an "authentic turpentine spirit" was and how one could define it. Some simply considered it impossible to establish a universal standard for all types of turpentine; for instance, the Austrian standardisation organisation pushed to construct a separate set of

[286] Departmental Archives of Gironde, Pine Institute Archives, box "ISO", file "Haye". "Essence de térébenthine, Projet de Recommandation ISO n. 67".

[287] Departmental Archives of Gironde, Pine Institute Archives, box "ISO", file "Haye". Letter from the Dutch normalization agency to AFNOR, 17 December 1953.

standards for each geographical region in order to differentiate, for example, turpentine from Bordeaux and turpentine from Austria, but in the end, the proposal was rejected.[288]

In 1955, the French delegation, represented by Brus, suffered yet another defeat. France was the only country that opposed the merging of two standards, one concerning gum spirits of turpentine and the other dealing with wood turpentine spirits. France tried to convince other committee members that "it is necessary ... to have as many standards as different chemical products recognised as such, independently of all business and economic considerations ... Gum spirit of turpentine is a well-defined product and it is necessary to regulate it with one specific norm".[289] As we can see, when the situation grew difficult for the French delegation, the former ambiguous distinction between trade and scientific definitions was quickly forgotten. The standards were not supposed to guarantee minimum quality of a commercial product but to reflect scientific reality. The choice of whether to create one or two standards was meant to be based on scientific considerations. Since the scientific expertise of the Pine Institute had confirmed that *essence de térébenthine* was a separate product, it had to be regulated on its own.

Paradoxically, the Dutch delegates criticised the French position in a 1955 letter to AFNOR, also by appealing to science:

> The working group is rather disappointed that the Secretariat of ISO/TC 35 [administrated by France] decided ... to adopt a position that does not take into account contemporary scientific achievements. Studies of the working group clearly prove that different types of turpentines form a unique entity. Their characteristics are

---

[288] Departmental Archives of Gironde, Pine Institute Archives, box "ISO", file "Haye 1953". Letter from Oesterreichischer Normenausschuss to AFNOR, 10 March 1953.

[289] Departmental Archives of Gironde, Pine Institute Archives, box "ISO", file "Haye". "Observations du Comité Membre Français sur le projet de recommandation ISO n. 364, 'Essence de Térébenthine et Essences de térébenthine de bois'", Document ISO/TC35 (Secrétariat-50) 82.

> identical for practical applications. ... The working group has the impression that AFNOR, instead of taking care of general interest, is prejudiced because France produces only gum spirits of turpentine ... The French definition [of turpentine spirit] comes from the times when other processes of extraction were not known ... when new processes are being developed, the language has to adapt ... The goal of ISO is to make things easier for chemists and trade agents. The new generation will not consider sufficient the historical fact that turpentine spirits had been once obtained from living pines, before other processes were invented, to justify publishing long specifications differing from one another only by a few numbers.[290]

While the descriptive power of science was used by the French to justify a diversity of definitions, the Dutch appealed to the "converging", "transformative" force of scientific progress. Both delegations enlisted scientific rhetoric to their advantage. It would probably be too far-fetched to claim that both of these positions reflect in reality some deeper cultural conditioning — that in France, a more descriptive, positivistic tradition of science made it a "stabilising" factor of post-revolutionary order, while in the Netherlands, science was more practical and engineering-oriented, thus reinforcing its dynamic aspect. It is, however, clear that these two narratives were used in practice by the arguing parties and that personal affiliation might have also played a role in constructing them. As a matter of fact, Brus, who represented France, was the head of a science faculty and a long-time scientist who contributed to numerous major discoveries concerning terpenes and pine resins, while his Dutch interlocutor, W. Biessel, was an engineer working for the industry.

ISO Standard 412 and After

In the following years, the Pine Institute continued to criticise the countries that opposed the French definition: "The determination of some

[290] Departmental Archives of Gironde, Pine Institute Archives, box "ISO", file "Haye". Letter from Dr Ing. W. Beissel from HCNN to AFNOR, September 1955.

Table 15: ISO Standard 412 (extract)

		Wood turpentine		
	Gum spirits of turpentine	Steam-distilled	Sulfate	Destructively distilled
Density 15.5 °C (Min–Max)	0.855–0.870		0.860–0.870	0.850–0.865
Refractive index		1.465–1.478		1.463–1.483
Distilled below 152 °C		0%		
Distilled below 170 °C		Min. 90%		Min. 60%
Polymerisation residue	Max. 12%		Max. 16%	

countries, not producing *essence de térébenthine* or producing *essence de bois*, to maintain ... the name turpentine for products that in reality try to replace it is a result of their own business interests tending to create confusion among customers."[291] Due to the French obstruction and because of the general change in business trends, work on the standard slowed down. That is why even though the first French draft of the standard was ready as early as 1953, the final standard, ISO 412, was only adopted 23 years later in 1976 (Table 15).[292]

The final version of the standard regulated two types of turpentine, gum spirits of turpentine and wood turpentine, the latter being divided into three categories inspired by the American regulations (steam-distilled turpentine, destructively distilled turpentine, and sulfate turpentine). It also stipulated that the density of gum spirit of turpentine should not be lower than 0.855 (rejecting the

[291] Departmental Archives of Gironde, Pine Institute Archives, box "ISO", file "Haye". "Observations du Comité Membre Français sur le projet de recommandation ISO n. 364 'Essence de Térébenthine et Essences de térébenthine de bois'", Document ISO/TC35 (Secrétariat-50) 82.

[292] Departmental Archives of Gironde, Pine Institute Archives, box "ISO". ISO Norm 412, "Gum spirit of turpentine and wood turpentines".

French-proposed minimum of 0.860), thus making it indistinguishable from steam-distilled wood turpentine in this regard. As a consequence, the only distinguishing factor for gum turpentine spirit was its polymerisation residue, which was slightly lower than for other turpentines. The only success of the French delegation was to maintain the traditional vocabulary in the French version of the standard, which meant that the term "essence de térébenthine" was clearly distinguished from "essence de bois" on the French market.

But this success was bitter. Already by 1968, Canada had revoked its own national standard in this field because of a "lack of use", and many countries followed. More and more sophisticated fractionation methods made it possible to obtain not a mixture but precisely defined terpenes that could be consequently used in more demanding industries. In 1987, in Budapest, it was proposed to revoke the relatively young standard 412, and in 1990, it was withdrawn by almost unanimous vote.

Conclusions: Scientific Definitions as Means of Business Pressure

The history of standardisation of turpentine demonstrates that the process of constructing the vocabulary to describe commodities is a phenomenon that is by no means confined to the transition period towards modernity in the eighteenth and nineteenth centuries. On the contrary, it continued throughout the twentieth century. The definitions coming from the *Encyclopédie* were almost completely abandoned, and the terminological borders were profoundly reshaped, not only because of subsequent scientific research but also mostly because the industrial vocabulary, limited to cheap pine resins, supplanted the first scientific attempts to define turpentine. Both the Pine Institute of Bordeaux and the USDA Bureau of Chemistry worked in service of the industry that used the name "turpentine" in a much more restrictive way than the authors of the *Encyclopédie* had. However, in spite of this general rejection of traditional categories, the researchers in France and the US in the 1920s and 1930s constructed definitions that were contradictory. They seemed to reflect national trade interests first and foremost; these were protectionist measures and not "objective" scientific categories. The process of negotiations surrounding turpentine continued after World War II. Discussions about turpentine continued after World War II, with experts worldwide vigorously debating the topic to find common ground, culminating in the publication of the ISO standard 412 on turpentine in 1976.

The meanings assigned to objects within specific economic, cultural, or natural settings can undergo redefinition when the context changes. The definitions of a commercially pure soap, a light rosin grade, or commercial gum turpentine inherently involve a certain degree of arbitrariness. That is why standardisation organisations and similar institutions play a critical role in dealing with these ambiguities, especially when multiple languages are involved. Contemporary specialists in linguistics and standardisation explain "natural and uncontrolled evolution of technical terms can no longer be relied on to ensure unambiguity and efficiency in use of language" [Teichmann, 2011, p. 531]. This means that the evolution of terminology should be guided by competent organisations, such as standardisation committees, that have sufficient resources to gather different stakeholders and build a consensus respecting trade interests of the relevant parties.

In the context of standards reliant on high scientific and technical expertise, such as in the case of turpentine, such an approach seems counterintuitive. Scientific definitions are intended to be independent of any business pressure. This is how French scientists defended their own trade standards, asserting that *essence de térébenthine* and *essence de bois* are simply two different chemical products and that this is a sufficient reason to produce two separate regulations. But at the same time, we clearly see that science did not have the final word in the negotiation process, at least not in a straightforward sense. The French specifications from 1931 did not radically differ from the ones in the ISO standard 412 of 1976. What evolved was the general perception of turpentine as the American approach, favouring practical applications, became prevalent. It was the industrial and economic context that changed, in the same way as it had changed in the nineteenth century, when pine resins dominated the market and, as a consequence, the old *Encyclopédie*'s categories became overridden. The industry designed the nomenclatural "moulds" that were later filled with scientific knowledge. Today, the "mould-shaping" procedure is formalised with the standardisation committees that constantly rework the vocabulary in use.

However, the history of standardisation of resinous products also highlights the significance of inertia in the evolution of trade terminology. Had France never developed a high-quality soap industry relying on sophisticated plant-based oils, the introduction of resin into soaps might not have become a problem. The Marseille soap industry, with its modern manufacturing plants, high degree of expertise, and, above all, long-term production traditions could exert influence on what should be considered a proper soap, and its perspective was deeply rooted in the conditions from the nineteenth century.

Sometimes, this inertia is almost imperceptible to the outsiders, as in the case of rosins. The Savannah rosin scale influenced the federal American scale, which was later imported to France. The marginal but still used French rosin scale has then its roots in the practices of nineteenth-century Georgia, but this genealogy was rendered invisible through successive translations of standards in different contexts.

The lesson to be learned from the standardisation of resinous products is that seemingly neutral standards can have significant economic consequences. In a good-will effort to ensure a consensus, stakeholders might be tempted to appeal to science as a common ground for standardisation. The problem, of course, is that science itself is not always free of bias. Having in mind the research gap in some fields between different countries, certain actors may profit from their privileged position by rejecting competing scientific mindsets that may impact standardisation. Only careful historical analysis of definitions can explain the role of science in setting standards. In this sense, historians and philosophers are as important as scientific experts in the process of constructing standards for complex substances whose meaning is contingent on social factors.

CONCLUSIONS: PINE RESIN CHEMISTRY AS GREEN CHEMISTRY *AVANT LA LETTRE*

"Like a terebinth I spread out my branches,
and my branches are glorious and graceful."
Sirach 24:16

The book demonstrated that the Pine Institute was a distinctive and multifaceted institution. It served as an applied research laboratory for the Landes forest industry, provided expertise in the standardisation of resin-based products, and gained international recognition as a centre for fundamental resin chemistry research. Throughout the chapters, we have explored these diverse facets and revealed the modular and eclectic nature of the Pine Institute. It was a hybrid organisation that connected and represented the interests of a wide array of stakeholders. This unique setting made pine resin chemistry, as a discipline, particularly conducive to industrial interests, not just by the means of funding but by adopting traditional industry-made trade categories. Resin chemistry was a particular case of industry-inspired university science. On the one hand, this meant that the entire pre-scientific messiness of resin terminology was incorporated into scientific frameworks of the discipline. On the other hand, it allowed the Pine Institute to acquire an unparalleled authority over the entire resin industrial sector, which could be exploited in times of crisis. For much of its history, resin chemistry was a "crisis science" promising to deliver the local economy from recurrent difficulties. It was the audacious promises of the French resin chemistry that made it particularly attractive to the American chemists who were eager to reproduce the French successes on their own soil. For that reason, even if the Americans surpassed the French and eventually even forgot the genealogy of their field, the history of resin chemistry is inextricably linked to the history of the Landes Forest and the Pine Institute.

Every single aspect of resin chemistry's history could have been isolated and studied under a different angle. It would be possible, for instance, to study the history of scientific ideas and techniques concerning resin and merely mention the institutions where these studies were conducted. It would be equally possible to write the history of the Landes resin industry and of its decline and include the Institute as one of its multiple stakeholders. This work juxtaposed these different dimensions to highlight their interdependencies. The book's main argument is that the scientific discipline of pine resin chemistry, as practised in the French and American laboratories in the first half of the twentieth century, was the fruit of unique social, economic, and environmental circumstances.

There were indeed countries, such as Spain, Portugal, Greece, Poland, and Russia, which possessed vast pine forests and were prominent producers and exporters of resinous products for decades but failed to develop full-blown institutions devoted to the chemistry of these raw materials. Likewise, despite being major consumers of resinous products, countries like England and Germany, with their well-endowed chemical laboratories, which worked on terpenes, resins, and their various components and applications, never fully institutionalised this knowledge.

This is not to say that these two categories of countries have not contributed to resin science. One could write a book entirely devoted to German contributions to the study of resins or to the technical innovations of the resin industry in Spain (which, incidentally, has been already published [Delgado, 2015]). But it was only in France and in the United States (US) where the worlds of chemical research and the pine resin industry connected successfully, and their relationship became formalised and institutionalised. In this sense, as noted in Chapter 2, the name "resin chemistry" can be misleading. First, because the discipline was centred specifically on pine resins and excluded a tremendous number of resins studied in various economic and geographic contexts. Second, because it was not just chemistry but an integrated continuum of research topics incorporating biology, forest management, economic analysis, harvesting techniques, and, last but not least, chemistry, strictly speaking. This "resin science" was, as demonstrated in Chapter 2, a prototype of what a sustainable science could look like before the term was coined.

The emergence of this eclectic discipline of resin science in France was due to the encounter of many concomitant factors. First, no resin chemistry would be possible if it was not for the forest of Landes — an artificially created and rationally managed wood and resin factory spanning over hundreds of square kilometres, designed to be sustainable and to provide for generations. This forest was large enough to sustain a huge industrial sector and give employment to tens of thousands of people. More importantly, the narratives concerning the forest were infused with the imaginary of progress and social improvement from its very inception, legitimising scientists and their contributions in the eyes of the forest's stakeholders. Second, whereas the resin industry was big and wealthy enough to fund research, it was also fragmented into multiple, often conflicted, stakeholder categories. These stakeholders were incapable of single-handedly subjugating scientists or enforcing one single research direction, but they were united in their interest in developing resin chemistry in general for the sake of the preservation of the existing economic system. This arrangement gave the Institute's researchers long-term financial stability but also sufficient autonomy to pursue projects that promised no immediate profit (*e.g.*, fractionation). Thanks to this situation, the Institute did not neglect basic theoretical research that would form the core element of the discipline. Finally, for the survival of an institution such as the Pine Institute to be possible, a pre-existing scientific environment was necessary. The Bordeaux Faculty of Sciences was an important scientific centre in the nineteenth century (for a French provincial university, at least) and maintained important traditions in agricultural research thanks to Ulysse Gayon. Therefore, it provided fertile ground for gifted chemists outside of Paris, such as Maurice Vèzes, Georges Dupont, and Georges Brus, to flourish in their careers and make themselves useful for the local industry.

Naturally, all these factors were closely knitted together and conditioned one by another. Their convergence led to an environment in which a unique self-perpetuating and sustainable knowledge production system could emerge. It could be argued that the American system, in which the initial conditions were different, proves that there was nothing unique about French resin chemistry. However, the American resin chemistry flourished because the Americans explicitly emulated the French solutions. They held as exemplary the French harvesting methods, forest management practices, and scientific institutions. They painstakingly introduced them one by one, and underlined their French parentage (be it Herty's cup and gutter system, the forestry policies advocated by the US Senate in 1920, or the Pine Institute of America) every single time. Even after their introduction, with an industry three times larger than the French one, the Americans still lagged far behind in theoretical foundations. The American resin chemistry was made possible because it was preceded by the French resin science, which had emerged organically, and only the dramatic events of World War II and the massive federal investment in agricultural research managed to change the relative position of the two countries in terms of scientific superiority.

As successful as it was in France in the first half of the twentieth century, the unique self-contained scientific-industrial system of the Landes Forest had its drawbacks. Due to its relative closure on itself, French pine resin chemistry created a specific vocabulary that could not be contested from the inside. The limitations of this vocabulary became only apparent when confronted with alternatives developed in other systems. The prime example of this conflict was, of course, the international standardisation of turpentine that demonstrated that making the same institution responsible for the basic research, modernisation of the industry, and standardisation of products on the national level, may have led to a certain myopia.

Becoming aware of these limitations is crucial if one wants to construct more sustainable knowledge production patterns in the future. The Pine Institute and resin chemistry, despite their ultimate disappearance, may still be considered a success story. The unique Landes scientific and industrial system remained in a state of healthy equilibrium until it was disrupted by external factors. On the one hand, the sustainable resinous products were replaced by cheaper yet unsustainable wood ones, and especially by equally unsustainable petroleum derivatives. On the other hand, the French and American gum naval stores industry failed because the same products could be harvested and imported from countries with lower labour costs. However, in both cases, what can be seen as economic victories against an obsolete and inefficient production system came at a heavy cost. The most obvious observation is that when sustainable feedstocks compete with unsustainable ones, the victory of the latter will always be Pyrrhic. But the problem goes deeper. In fact, the departure of naval-stores-producing industries towards the global south led to the disappearance (or reconfiguration) of resin chemistry institutions. As a consequence, a tremendous amount of original know-how, developed and practised for decades, was lost. In this new setting, the resin production industry, now based in developing countries, was divorced from the consuming one. But it was precisely the superposition of these two dimensions, downstream and upstream, that made resin chemistry an autonomous and successful discipline for such a long time.

Interestingly, the study of the history of resin chemistry has demonstrated that it was a form of green chemistry long before the term entered into the public debate. Resin chemistry was not green simply because it dealt with a natural renewable resource such as resin, although the renewable character of the raw material is also important. For instance, it is noteworthy how little transformation resin needs to become a marketable product. A simple distillation separating rosin and turpentine can provide a range of substances directly usable as solvents, fuels, waxes, or in paints and varnishes. Compare this to white spirit, a product that requires breaking complex molecular chains in high-pressure, high-temperature sophisticated refineries. Much of the innovation in resin chemistry relied on preserving the complexity of the natural substances by changing them as little as possible. However, even this feature was not enough to consider a given branch to be fully sustainable.

One of the precepts of modern green and sustainable chemistry is that it should take into account the entire life-cycle of a given product [Jessop, 2020]. Its toxicity and environmental impact should be evaluated starting from the extraction, through its transformation in production plants, all the way to its use by the consumers. But is such a theoretical *modus operandi* possible from the practical standpoint if raw materials are extracted by cheap labour hundreds of miles away from the intermediate-industrial plants, which are themselves far away from the consumers? The problem is not, of course, exclusively due to geographical distance (although the importance of physical separation is not to be underestimated). The key issue here lies in the comprehensive involvement of all the industry's stakeholders in the process of knowledge production and dissemination. Sustainable chemistry, with its holistic ambitions, requires a multilevel feedback and knowledge flow. The Pine Institute exemplified this aspiration by actively engaging with distilleries and forest workers while also providing expertise to consuming industries. Its scientists possessed a deep understanding of the challenges at every stage of the products' life cycle.

The construction of a comprehensive "green" knowledge production system had challenges of its own. Every single technological advancement had to be negotiated with a broad range of stakeholders. Instead of embracing radical innovations, the Landes industry and its research institutions adopted an iterative approach focused on small improvements. From an outsider's viewpoint, this might seem conservative, but over several decades, these incremental steps led to remarkable innovation and the generation of unique knowledge that could only emerge within this specific context.

However, the Landes socio-economic model did not align well with the neoliberal, exploitative, and unsustainable economic system that gradually dominated the world in the second half of the twentieth century. Today, it is rare to find a university-based (yet largely autonomous), industry-oriented (yet labour-friendly), and environmentally proactive (yet multidimensional in its analyses) institution like the Pine Institute. Nonetheless, the challenges of the climate crisis, the inevitable peak oil, and the global geo-political power shifts may compel us to reconsider the value of smaller, local, and sustainable industries. In this future setting, a novel way of organising knowledge production — a new, not merely chemistry, but an entire "green science" — has to be (re)created. Past experiences, especially from the first half of the twentieth century, can guide us in making it happen and avoiding the pitfalls faced by these early pioneering institutions.

Annex

TEACHER'S GUIDE TO GREEN CHEMISTRY *AVANT LA LETTRE*: THE PINE INSTITUTE AND RESIN CHEMISTRY IN AQUITAINE (1900–1970)

The history of the Pine Institute and resin chemistry is rich and multilayered. It can serve as an introduction to a wide range of issues for students at different levels of education. This didactic annex aims at helping teachers and instructors navigate the book and identify problems that might be potentially incorporated into their classes. Here, I present three lessons that can be explored in the classroom based on the content of the book, each of them divided into three tasks with appropriate readings, suggested activities, and key learnings. The first one targets high-school-level students in classes such as general science, chemistry, natural history, and geography and uses the case of resin chemistry as an introduction to sustainability studies. The second lesson targets university-level chemistry students and is meant to incite young chemists to reflect on the challenges of their own discipline as well as to introduce some elements of epistemology. The third lesson is for Science and Technology Studies (STS) or history of science students and raises overarching questions about knowledge production environments.

These lessons demonstrate the versatility of the book's content and provide a foundation for educators to use it in their teaching. By exploring these lessons, students can gain a deeper understanding of sustainability, epistemology, and the history of science while also developing critical thinking skills and a broader perspective on the role of science and technology in society.

The themes presented in the book are not necessarily self-contained in the sense that they may require readings from outside the book. The book itself, however, can constitute a starting point for many and much more general and contemporary debates with which the students could potentially connect. Teachers are invited to identify relevant contemporary debates in their countries or regions that can be studied side by side with the Pine Institute's story.

Different lessons have different target groups, but the division is not strict. Although Lesson 1 is probably the only one that might be suitable for high-school students, it can also form a basis for classes aimed at more advanced students, regardless of disciplinary background. Similarly, Lesson 2 targets chemistry students specifically, but it may also be of interest to general history and sociology of science students. Lesson 3 is, in principle, designed for students with more familiarity with the methodology of social sciences but can be adapted to different contexts.

Each lesson is divided into three study problems that refer to specific fragments of the book. Every task consists of the following components:

- "Readings" consist of excerpts from the book and external source materials that students are encouraged to review before attending the class. Instructors have the flexibility to modify or refine these suggestions according to their preferences or the time available to the students.
- "Activity" is both a commentary and a step beyond the source materials, and students are expected to work on it during class or at home.

- "Key learning" is based, first and foremost, on the text material but can be enriched by the outcomes of the activity.
- "General discussion" is intended to go beyond the problems presented in the readings and discussed in class.

Needless to say, all of these are mere pointers and indications and do not replace informed guidance from the instructors who need to adapt the materials to their students and to different contexts. Depending on the class format, a single problem/task/case can be sufficient for a two-hour-long class, if an appropriate contemporary case study is selected for comparison. Additionally, teachers are invited to combine study problems from different lessons if they find them relevant for a specific group of students they work with.

Finally, this annex is, above all, a toolbox that can be applied in a wide range of circumstances to introduce elements of the history of science.

Lesson 1

Title: Sustainability and environment: A case of the pine resin industry

Target audience: High school students taking chemistry, natural history, geography, or any class related to sustainability.

Purpose: The aim of the lesson is to introduce students to the notion of sustainability and renewability of raw materials. It is also an opportunity to learn about the history of the forest industry in general and about the influence exerted by people on the environment to better understand the impact human activity has on our planet.

1. Problem: Are forests natural?

Pines have been planted and employed in many countries across the globe (US, France, Germany, Mexico, Spain, Greece, Portugal, Scandinavian countries, Eastern Europe...). This makes pine forests relatable to students in different countries and in different geographical zones. However, they may not be aware of: (1) the artificial character of many pine plantations, (2) the economic motivations behind their establishment, and (3) the significance of the industries they spurred. The book's accounts of French and American forests can serve as a gateway to understanding environmental history more broadly.

Readings:

- Chapter 1 — Subchapter 1. Economic Prehistory of the Institute: the Landes Forest in the Nineteenth Century
- Chapter 5 — Subchapter 1. The Origins of the American Naval Stores Industry
- For France, Wikipedia article on the Landes Forest: https://en.wikipedia.org/wiki/Landes_forest
- For the US, website of the American Forests association, on the history of longleaf pines: https://www.americanforests.org/blog/history-of-the-longleaf-pine/

Activity: Discuss: How do the histories of the French and American forests differ? Look at the economic history of your country/region online. Are there any pine forests? Are they natural or artificial? Did their existence link to industrial activities in your region? If not, try to find other similar forest plantations that have been used for industrial reasons.

Key learnings: Students can explain how concepts as distinct as the environment, technology and economy are interconnected and that the environment is not always "natural" in the sense that it does not pre-date human intervention.

2. Problem: Is natural always greener?

Pinewood derivatives have a long history of use, and their applications have evolved over hundreds of years. Initially, they were used as "naval stores" and served as raw materials for the manufacture of tars, charcoal, essential oils, soaps, and many other products. However, with the onset of the second industrial revolution, new and more sophisticated uses were developed thanks to synthetic chemistry. Many of these new resin-based products competed with other natural products, such as camphor. Natural camphor was extracted from camphor trees in Taiwan in a process that was extremely destructive to the local forests. Synthetic camphor, on the other hand, was a complex and wasteful industrial process relying on renewable turpentine.

Readings:

- Turpentine (Wikipedia, https://en.wikipedia.org/wiki/Turpentine) and camphor (Wikipedia, https://en.wikipedia.org/wiki/Camphor)
- Asianometry's video on camphor: White Gold: How the West Broke Taiwan's Camphor Monopoly (Youtube video, https://www.youtube.com/watch?v=VxWt36579Hg)
- Chapter 2 — Subchapter 4. Turpentine Spirit, in Synthetic Camphor: a New Hope for the Landes Industry

Activity: In the case of camphor, the synthetic turpentine-based product competed with an identical natural one. Discuss in small groups: Was natural camphor production greener and more sustainable than the synthetic one? Think about different raw materials that have been used since antiquity and still remain in use and how they were transformed by scientific and technological advancements. Is the use of the plant raw materials always good for the environment? What does it mean that a resource is renewable?

Key learning: Students can illustrate why "greenness" is not intrinsic to a given product or a process. They should be able to describe that an entire production system and the product's lifetime has to be evaluated to assess sustainability.

3. Problem: Can we imagine a civilisation without petroleum?

Many pine wood products competed with petroleum ones; *e.g.*, turpentine with white spirit, natural with synthetic resins, wood gas with petrol, *etc.* Many forest owners saw themselves as direct competitors of oil potentates. Some of these applications were spectacular;

for example, turpentine was used for the manufacture of rocket fuels in the French space programme.

Readings:

- Google "turpentine fuel" and see what you find!
- Chapter 3 — Subchapter 5. Resins and Space Conquest: Véronique rockets

Activity: Think-pair-share: What petroleum substitutes have you heard about? Do the vehicles in your country/region rely on fossil fuels? Are petrol substitutes always greener?

Key learning: Students are expected to take away from the class that even in high-tech applications such as rocket science, plant-based renewable substances can successfully substitute petroleum. This helps to dispel the notion that phasing out fossil fuels will halt progress and innovation.

Class discussion: Discuss the tension between the petroleum and non-petroleum-based industries. What are the challenges of the transition from an economy based on fossil fuels towards the one relying exclusively or mostly on renewable feedstocks? Do you believe that a global shift to forest products would be sustainable today? What is the place of chemistry in the debate on renewable materials and energies?

Lesson 2

Title: Pine chemistry: A discipline of the past in service of challenges of the future

Target audience: University-level chemistry students

Purpose: The aim of this lesson is to introduce future practitioners of chemistry to the challenges encountered by their predecessors. The lesson should incite some discussions on the historical process of knowledge accumulation and especially on the plurality of roles of chemists in society.

1. Problem: How can the composition of resin be determined?
Pine resin is a mixture of terpenes and resin acids. To separate these components, they are distilled. However, distillation involves heating that can alter the substances in question. Moreover, the distillation products are rarely pure. As a result, theories concerning the composition of pine resin have varied significantly over the years. Every now and then, chemists discover new types of resinic acids only to discard their existence later. It took decades before the full consequences of thermal transformation were fully understood.

Readings:

- Chapter 2 — Subchapter 1. What is Resin?
- Chapter 2 — Subchapter 3. Between the Laboratory and the Forest: Contributions to Pine Gum Treatment
- Chapter 2 — Subchapter 5. Rosin/Colophony
- https://www.chemspider.com/

Activity: Look for the structural formulas of major resin terpenes and acids in contemporary articles, chemistry manuals, or chemical databases. What are their chemical and physical properties? Why do you think the identification of resin acids was so problematic for chemists?

Key learnings: Students should be able to recognise the complexity of many natural organic substances and how their compounds may vary depending on the geographical origin or the extraction process. Based on the history of rosin research, they should have understood that the identification of compounds in a given material, which may be casually listed in a chemistry manual today, is often the fruit of decades of work of hundreds of scientists all over the world.

2. Problem: What is the role of chemists in creating trade standards?
While the core components of pine turpentine are usually pinenes, their proportions vary, and some impurities are almost always present. However, it is impossible to determine the substance's exact composition without laboratory equipment. Therefore, scientists were entrusted with the mission of standardising turpentine by

establishing criteria for everyday use by the industry and consumers. Over time, different standards emerged, often conflicting with one another.

Readings:

- Chapter 6 — Subchapter 2. Defining Turpentine Spirit in France
- Chapter 6 — Subchapter 3. International Standardisation of Turpentine (1950–1976)
- Hepler-Smith, E. (2015). Systematic Flexibility and the History of the IUPAC Nomenclature of Organic Chemistry, *Chemistry International*, 37, 2, pp. 10–14.

Activity: Discuss in groups: Think about the challenges industrial chemists have to face when it comes to defining the manufactured products. How different are trade standards and scientific definitions? What are their respective goals? Reflect on the evolution of naming conventions in chemistry and the relation between systematic names and common names. Are systematic names always more practical from the industrial point of view?

Key learning: Students should be able to describe how scientists standardising turpentine represented the economic interests of their countries using the authority of science to defend their definitions. Students should use the problem described in this lesson to explain and exemplify how science can be used as a rhetorical tool for business pressure.

3. Problem: One man's waste is another man's treasure: how do chemists adapt to novel circumstances?

The differentiation between what is considered waste and commercially interesting raw material is not fixed once and for all. Chemists often find new and creative ways of extracting value from seemingly useless objects by bridging unrelated fields of activity. This was the history of the American wood naval stores industry. A large chemical conglomerate, the Hercules Powder Company, helped to extract dead pine stumps from the ground using explosives, only to discover that pine stumps could be a valuable raw material on their own.

Readings:

- Chapter 5 — Subchapter 3. Reclaiming the Wastes and the Discovery of Wood Naval Stores

Activity: Think-Pair-Share: Think about the role of creativity in scientific research. Can you recall instances where waste materials were turned into a profitable business through innovative methods? In your opinion, can the wood naval stores industry be considered "green" despite not being renewable?

Key learnings:

(1) Students should be able to explain how the raw materials or feedstocks are not always natural, and provide examples of man-made materials.
(2) Students should be able to discuss and exemplify how innovation in science can come from the most unusual encounters between different fields of knowledge and different branches of industry.

Class discussion:

(1) What are the similarities and differences between the treatment and exploitation of biomass and petroleum? Focus on terpenes and terpenic acids (think about modern biorefineries).
(2) Think about the role of chemists not as dispassionate and neutral analysts, but as stakeholders in economic battles.

Lesson 3

Title: Understanding knowledge-producing environments (case of resin chemistry)

Target audience: STS/history of science and technology students (Master/Ph.D.) The lesson can take varied formats: seminar, group presentation, essay, *etc*.

Purpose: This lesson targets future academics, teachers, and science communicators. It introduces students to a wide range of challenges concerning the study of knowledge production environments and their complexity. Since the audience consists of people familiar with social sciences and humanities, they are expected to read ample fragments of the book.

1. Problem: How is science funded?

The Pine Institute was funded by labour unions, private companies, the national government, and local public authorities. Interestingly, its prosperity seems inversely correlated to the prosperity of the local resin industry. It is only when the latter is deteriorating that investment flows to the Institute. To put it differently, the Institute sold "hope" that through science, prosperity would come to the entire community.

Readings:

- Chapter 4 (all): Can Science Feed on a Crisis: The Relations Between the Pine Institute and its Economic Environment
- Wikipedia: Knowledge production modes (https://en.wikipedia.org/wiki/Knowledge_production_modes)
- For a more complete overview, see: Etzkowitz, H. and Leydesdorff, L. (2000). The Dynamics of Innovation: From National Systems and "Mode 2" to a Triple Helix of University–Industry–Government Relations, *Research Policy*, 29, pp. 109–123

Activity: *Group Discussion*: As preparation for discussion, study concepts such as "Mode 2" or "Triple Helix" and consider whether the Pine Institute fits into these theoretical frameworks. How do you think the relation between public research institutions and the world of industry should look like?

Key learnings: Students are able to explore how scientists and scientific institutions in a capitalist economy need to constantly justify their expenses and activities by employing the outdated language of scientific progress presented as the solution to problems that remain inherently social and economic in nature. Students should be able to describe the mismatch that arises between techno-scientific expectations and real research as a consequence of the false perception of what science is capable of and how this mismatch undermines public trust in science.

2. Problem: How does knowledge circulate between different countries?

The French used to be leaders in the field of (applied) resin chemistry, with their books widely read on the other side of the ocean. The Americans were lagging behind and tried to catch up throughout the 1910s, 1920s and 1930s. After World War II, the situation changed drastically. The Americans became the leaders in the field, and it was the French who had to follow.

Readings:

- Chapter 5 — Subchapter 1. The Origins of the American Naval Stores Industry
- Chapter 5 — Subchapter 2. American Resin Chemistry Research Infrastructure Before World War II
- Chapter 5 — Subchapter 4. Gum Naval Stores During and After World War II

Activity: Think-Pair-Share: Based on the texts, identify the reasons behind America's backwardness, notably in relation to their conservation policies. Identify the reasons for the shift in power relations that made them leaders in the field after World War II. Examine which factors were unique to resin chemistry and which factors illustrate broader developments during that period.

Key learnings: Students are able to understand how knowledge circulation patterns between scientific institutions that work in favour of industries competing with each other are always complex and that these institutions themselves often have to navigate between collaboration and competition. Additionally, they should understand the tension between the universality of science as a collective endeavour and the constraints coming from market economy.

Problem 3: What can we learn from scientometrics?
Scientometrics is the art of measuring the productivity of scientists, but it can also yield valuable insights into the history and development of scientific disciplines and institutions. By examining publication patterns of the Pine Institute, it is possible to uncover interesting trends in the field of resin chemistry from the 1920s to the 1960s.

Readings:

- Chapter 1 — Subchapter 3. The Pine Institute Bulletin
- Chapter 3 — Subchapter 2. Institutional Architecture of the Laboratories of the Pine Institute and their Scientific Production after the War (1948–1969)

Activity: Group Discussion: Discuss the similarities and differences between the two scientometric analyses. What do they tell us about the institution and the discipline they represented? What are the major trends? Was there anything that surprised you about this data?

Key Learning: Students can describe how different levels of analysis reveal different tendencies and explain why there will also be blind spots in scientometric analyses; the scientist is not merely a machine to produce articles. Students can give examples of the scientists' roles in disseminating and communicating scientific results as well as in science management; these activities are easy to miss when we view science exclusively through the lens of quantitative tools.

Final Task/General Discussion: Explore the big themes studied in all previous lessons.

(1) Sustainability in resin chemistry: We have learned about the historical and contemporary uses of resin as a raw material in the chemical industry. As we confront the demands of a post-petroleum era, it becomes pertinent to examine how resin chemistry can play a role in facilitating the transition towards a more sustainable economy. Does resin possess the potential to

be a sustainable raw material of choice, or does its production carry adverse environmental consequences? Moreover, can we adapt and implement the institutional framework of resin chemistry that was prevalent in early twentieth-century France to the modern chemical industry?

(2) Interconnectedness of knowledge production, economy, and environment: We have observed the deep interconnection between scientific institutions, economic factors, and the natural environment. Is it feasible to disentangle knowledge production from its economic context? How can we effectively strike a balance between economic development and environmental sustainability?

(3) The role of STS/HTS in sustainability debates: Throughout these lessons, we have engaged with the Science and Technology Studies (STS) and History of Technology and Science (HTS) approaches to understand the challenges of sustainability in the chemical industry. How can STS/HTS scholars contribute to contemporary debates on sustainability, and what strategies should they adopt to engage with chemists and engineers not only in academia but also in the chemical industry?

GLOSSARY

A

AFNOR: French Standardization Association (*Association française de normalisation*).

B

Brais: French term to describe dark coloured rosin.

C

Carre: French term for the "face" on the pine. See: Face
Colophane: French term to describe light coloured rosin.

D

D.R.T.: *Dérivés Résiniques et Terpéniques*, the largest resin producing company in the Landes; one of the few that has survived until today.

F

Face: the wound on the pine from which flows the resin.

G

Galipot: French term for the solid suspension in resin and dried rosin on the tree.

Gum: the traditional term for pine resin, although incorrect from the botanical standpoint.

L

Light wood: pine wood heavily saturated with resin.

L.R.B.A.: *Laboratoire de recherches balistiques et aérodynamiques* (Laboratory of Ballistics and Aerodynamics Research).

N

Naval stores: English term for pine resin derivatives and final products.

N.U.C.R.: *Nouvelle Union Corporative des Résineux*. A powerful resin trade trust created to replace U.C.R. in the late 1940s.

R

Rosin grade: a way for measuring the colouration of solid products of resin distillation.

S

SOLARED: *SOciété LAndaise des Résines Et de leurs Dérivés*, a trade company established in the 1950s by the biggest resin business stakeholders in the Landes to renew the local industry.

T

Térébenthinage: The French name for the process of cleaning pine resin (or transforming it into *térébenthine*). Introduced in the United States on a massive scale only in the twentieth century.

Terpenes: a category of hydrocarbons present in many plants, in particular, conifers, known for their specific aroma.

U

U.C.R.: *Union Corporative des Résineux*, a resin trade trust established in the Landes in 1939.

W

White spirit: the most popular petroleum-derived solvent.

FURTHER READING

On Resins and their Uses

Langenheim, J. (2003). *Plant Resins: Chemistry, Evolution, Ecology and Ethnobotany* (Timber Press, Portland).

Panda H. (2008). *Handbook on Oleoresin and Pine Chemicals (Rosin, Terpene Derivatives, Tall Oil, Resin & Dimer Acids)* (Asia Pacific Business Press, Delhi).

Zinkel, D. F. (1989). *Naval Stores, Production, Chemistry, Utilization* (Pulp Chemical Association, New York).

On the History of Naval Stores

Delgado, J. L. (2015). Del bosque a la fábrica: Técnica y ciencia de la resina de pino en la España contemporánea (Ph.D. Dissertation, Autonomous University of Madrid).

Delgado, J. L. (2016). Industrialization and Landscape: Gum Resin Pinewoods of France, Spain and United States in the 19th and 20th Centuries, *Estudios Rurales*, 6 (11), pp. 48–69.

Outland, R. (2004). *Tapping the Pines* (Louisiana State University Press, Baton Rouge).

Prizer, T. C. (2009). Pining for turpentine: Critical nostalgia, memory, and commemorative expression in the wake of industrial decline (Master dissertation, University of North Carolina).

Sargos, J. (1997). *Histoire de la forêt landaise: du désert à l'âge d'or* (L'Horizon chimérique, Bordeaux).

On the History of French Universities in the Provinces

Nye, M. J. (1986). *Science in the Provinces: Scientific Communities and Provincial Leadership in France, 1860–1930* (University of California Press, Los Angeles).

Paul, H. W. (1985). *From Knowledge to Power: The Rise of the Science Empire in France, 1860–1939* (Cambridge University Press, Cambridge).

On Green Chemistry

Anastas, P. T. and Warner, J. C. (1998). *Green Chemistry: Theory and Practice* (Oxford University Press, Oxford).

Krasnodębski, M. (2022). *Green Chemistry: A Brief Historical Critique* (Peter Lang, Berlin).

Linthortst, J. A. (2023). *Research between Science, Society and Politics. The History and Scientific Development of Green Chemistry* (Eburon Academic Publishers, Utrecht).

On Plant-based Alternatives to Fossil Economy in the Early 20th Century

Ekerholm H. (2012). Cultural meanings of wood gas as automobile fuel in Sweden, 1930–1945, in *Past and Present Energy Societies*, eds. Möllers, N. and Zachmann, K. (transcript-Verlag, Bielefeld), pp. 223–248.

Krasnodębski M. (2021). Upscaling Forest Waste: The French Quest For Fuel Autarky after World War I. *Technology and Culture,* 62, pp. 105–127.

Manuel J. T. (2021). Lessons from a Forgotten Fuel: Assessing the Long History of Alcohol Fuel Advocacy and Use in the United States, *History and Technology*, 37, pp. 411–428

Uekötter F. (2021). The Revolt of the Chemists: Biofuels, Agricultural Overproduction, and the Chemurgy Movement in New Deal America, *History and Technology*, 37, pp. 429–445.

BIBLIOGRAPHY

Alquier, B. (2009). *Le soulèvement des résiniers landais en 1907* (EDR, Cressé).

Anastas, P. T. and Warner, J. C. (1998). *Green Chemistry: Theory and Practice* (Oxford University Press, Oxford).

André, G. (1930). Le retour d'Amérique, *Bois et Résineux*, 6 April 1930.

Antoniotti, S. (2013). *Chimie verte — Chimie durable* (Ellipses, Paris).

Antunes, S. (2012). *DIY Satellite Platforms* (O'Reilly Media, Inc., Sebastopol).

Aufan, R. and Thierry, F. (1990). *Histoire des produits résineux landais* (Société Historique et Archéologique d'Arcachon et du Pays de Buch, Arcachon).

Averitt, J. A. (2007). *Families of Southeastern Georgia* (Genealogical Publishing Company, Clearfield).

Barraud, M. (1924). Rapport du Sénat Américain, *Bulletin de l'Institut du Pin*, 1, no page.

Barraud, M. (1927). La Récolte et le Traitement de la gemme: cours de M. le professeur Dupont, *Bulletin de l'Institut du Pin*, 35, pp. 79–85.

Barraud, M. (1927b). Contribution à l'étude des essences de térébenthines (Ph.D. dissertation, Faculty of Sciences of Bordeaux).

Barraud, M. (1930). Défense de l'essence de térébenthine contre la fraude et les succédanés étrangers, *Bulletin de l'Institut du Pin*, 7, pp. 145–150.

Barraud, M. (1930b). Historique du développement économique des Landes, *Bulletin de l'Institut du Pin*, 12, pp. 265–270.

Barraud, M. (1931). La Gemme et ses constituants, *Bulletin de l'Institut du Pin*, 23, pp. 245–250.

Barraud, M. (1931b). Historique du développement économique des Landes. *Bulletin de l'Institut du Pin*, 13, pp. 1–8.

Barraud, M. (1932). L'utilisation des produits résineux, *Bulletin de l'Institut du Pin,* 31, pp. 155–159.

Barraud, M. (1933). Les qualités marchandes de la gemme, de l'essence de térébenthine et de la colophane, *Bulletin de l'Institut du Pin*, 43, pp. 129–136.

Beach, E. S. and Kundu, S. (2017). *Tools for Green Chemistry* (Wiley-VCH, New York).

Beauclair-Lafaye, C. (1914). Recherche de la fraude dans la gemme et l'essence de térébenthine (Ph.D. dissertation, Faculty of Sciences of Bordeaux).

Bellini delle Stelle, U. (1929). Procédé et dispositif pour l'extraction de la gomme, de la résine et des essences des plantes en général, patent publication date: 23 April 1929, publication number: FR655733 A.

Bellini delle Stelle, U. (1931). Procédé de récolte des produits d'élaboration et de sécrétion des plantes, patent publication date: 7 January 1931, patent publication number: FR41359 E.

Berzelius, J. J. (1831). *Traité de chimie minérale, végétale et animale, 2e partie — chimie organique*, volume five (Imprimerie de Firmin Didot Frères, Paris).

Bijker, W. E. (1987). The social construction of Bakelite: Toward a theory of invention, in *The Social Construction of Technological Systems: New Directions in the Sociology and History of Technology*, eds. Bijker, W. E., Hughes, T. P. and Pinch, T. J. (MIT Press, Cambridge), pp. 159–187.

Blain, H. (1925). L'Huile de Bois et la Colophane dans l'industrie des peintures et vernis, *Bulletin de l'Institut du Pin*, 18, p. 256.

Blancheton, B. (1997). La stabilisation Poincaré est-elle à l'origine d'une récession? Les malentendus d'une controverse, *Histoire, économie et société*, 4, pp. 709–741.

Bodin, A. (1927). Les résineux devant le progrès scientifique, *Bois et Résineux*, 22 May 1927.

Bodin, A. (1928). Les savons résineux: la Chambre a voté le projet, *Bois et Résineux*, 25 March 1928.

Bonnin, A. (1950). Rapport moral de la Nouvelle Union Corporative des Résineux sur la campagne 1949–1950, *Bois et Résineux*, 30 September 1950.

Borglin, J. (1939). *Treatise on Turpentine and Pine Oil* (Hercules Experiment Station, Wilmington).

Borup, M., Brown, N., Konrad, K. and van Lente, H. (2006). The Sociology of Expectations in Science and Technology, *Technology Analysis & Strategic Management*, 18(3–4), pp. 285–298.

Bouyssou, L. (1931). Une interpellation de M. Léo Bouyssou sur la crise des Produits Résineux, *Bois et Résineux*, 8 March 1931.

Bradwell, J. (1947). French Naval Stores Experts See Valdosta, Tour Gum Belt, *The Valdosta Daily Times*, 42(140), 29 March 1947, p. 1.

Brodschi, J (1938). *Contribution à l'étude du cracking de la colophane* (Fredou et Manville, Bordeaux).

Brus, G. (1929–1930). Recherches sur le pinène et le nopinène, *Bulletin de l'Institut du Pin*, n. 65, 1929 – n. 9, 1930.

Brus, G. (1934). Sur l'Industrie du Camphre synthétique en France (conférence faite le 25 juin 1934 à la Journée des Résineux), *Bulletin de l'Institut du Pin*, 54–55, pp. 105–113.

Brus, G. (1935). L'Utilisation des produits résineux, *Bulletin de l'Institut du Pin,* 10, pp. 193–201.

Brus, G. (1935b). Rapport sur le fonctionnement et les travaux de l'Institut du Pin en 1934, *Bulletin de l'Institut du Pin*, 3, pp. 41–45.

Brus, G. (1936). L'Utilisation des produits résineux dans les revêtements routiers, *Bulletin de l'Institut du Pin*, 19, pp. 159–166.

Brus, G. (1937). Rapport sur le fonctionnement et les travaux de l'Institut du Pin en 1936, *Bulletin de l'Institut du Pin*, 27, pp. 63–67.

Brus, G. (1938). Les industries dérivées du pin, *Bulletin de l'Institut du Pin*, 47, pp. 233–241.

Brus, G. (1940). Les produits résineux et leurs emplois, *Sud-Ouest Economique*, December 1940, pp. 5–8.

Brus, G. and Barraud, M. (1935). L'utilisation des Produits Résineux dans les revêtements pour routes, *Bulletin de l'Institut du Pin,* 1, pp. 11–16.

Brus, G. and Barraud, M. (1937). L'utilisation des produits résineux comme insecticides et comme mouillant en agriculture et particulièrement en viticulture, *Bulletin de l'Institut du Pin*, 29–30, pp. 108–118.

Brus, G. and Legendre, P. (1947). Hydrogénation des acides résiniques et des produits qui en contiennent et qui en dérivent, patent publication date: 24 June 1947, patent publication number: FR922259.

Brus, G. and Legendre, P. (1951). P. Nouveau mode de traitement des gemmes de conifères, patent publication date: 27 September 1951, patent number: FR990892.

Brus, G. and Legendre, P. (1951b). Procédé de fractionnement des colophanes et brais de résine, patent publication date: 27 September 1951, patent number: FR990893.

Brus, G. and Legendre, P. (1956). Essence de térébenthine et dérivés, in *Chimie des Peintures, Vernis et Pigments*, volume II, eds. Champetier, G. and Rabaté H. (Dunod, Paris) [extract in Departmental Archives of Gironde, Pine Institute Archives, box "Divers"].

Brus, G. and Menaut, D. (1938) L'Institut du Pin: Ses Buts — Son organisation, in *Rapports et communications sur le pin maritime et ses dérivés présentés au 62e congrès tenu à Arcachon du 22 au 27 septembre 1938* (Association française pour l'avancement des sciences, Bordeaux), pp. 85–93.

Brus, G., Lê Van Thoi and Françoise, M. (1953). L'acide di-abiétique et ses dérivés, *Peintures, Pigments, Vernis*, 29(1) [extract in Departmental Archives of Gironde, Pine Institute Archives, box "Divers"].

Brus, G. Legendre, P. and Niolle, M. (1952). Nouveau mode de fractionnement des gemmes de conifères, *Peintures Pigments Vernis*, 28(10). [extract in Departmental Archives of Gironde, Pine Institute Archives, box "Divers"].

Brus, G., Lê Van Thoi, François, H. and Fines, C. (1952). Nouveaux dérivés de l'acide maléo abiétique, *Peintures, Pigments, Vernis*, 28(12), pp. 865–870.

Buffault, M. (1932). Un nouveau procédé de gemmage du pin maritime, *Comptes rendus des séances de l'Académie d'agriculture de France*, 18(1), pp. 90–93.

Buonanni, F. (1731). Trattato sopra la vernice detta comunemente cinese (Per Antonio de Rossi, Rome).

Burnett, W. B. (1930). *Pine Institute Technical Bulletin No. 1: Rosin in Synthetic Resins* (Pine Institute Publications, Jacksonville).

Burnett, W. B. and Odell, I. H. Jr. (1927). *Pine Institute of America Abstracts*, 1 (1).

Bush, V. (1945). *Science The Endless Frontier* (United States Government Printing Office, Washington).

Capus, J. (1931). La Fraude sur l'essence de térébenthine, *Bois et Résineux*, 11 January 1931.

Carson, J. A. G. (1921). The increased cost of naval stores production, in *Naval Stores: History, Production, Distribution and Consumption*, ed. Gamble, T. (Review Publishing & Printing Company, Savannah), pp. 73–74.

Chaine, J. (1935). Discours prononcé aux Obsèques de M. le Professeur Vèzes, *Bulletin de l'Institut du Pin*, 2, p. 26.

Charle, C. (1989). Le Personnel dirigeant du CNRS (1937–1966), *Cahiers pour l'histoire du CNRS*, 4, pp. 1–30.

Courret, E. (1929). Perfectionnements aux appareils servant à la distillation des résines ou autres matières, patent publication date: 4 October 1929, patent publication number: FR666709 A.

Courret E. (1934). Perfectionnements apportés aux procédés et appareils pour le traitement des gemmes de pin et des produits susceptibles d'en être extraits, patent publication date: 21 March 1934, patent publication number: FR761546 A.

Courret, E. (1940). Perfectionnements apportés aux procédés et dispositifs pour la distillation, en marche continue et dans le vide, des gemmes, notamment des gemmes de pin, patent publication date: 3 December 1940, patent publication number: FR858784 A.

Courtebras, B. (2008). *La structure des hiérarchies scolaires: histoire de l'organisation de l'enseignement professionnel et technique* (Publibook, Paris).

Croteau, R. (1998). The discovery of terpenes, in *Discoveries in Plant Biology vol. 1*, eds. Kung, S.D. and Yang, S. F. (World Scientific Publishing, London), pp. 321–343.

Crouch, C., Le Galès, P., Trigilia, C. and Voelzkow, H. (2001). *Local Production Systems in Europe. Rise or Demise?* (Oxford University Press, Oxford).

Dardel, J.-D. (2006). Le Conseil supérieur de la recherche scientifique et du progrès technique vu de l'intérieur, in *Le gouvernement de la recherche. Histoire d'un engagement politique, de Pierre Mendès France à Charles de Gaulle (1953–1969)*, eds. Chatriot, A. and Duclert, V. (La Découverte, Paris), pp. 205–211.

Daumas, J.-C. (2007). Dans la "boite noire" des districts industriels, in *Les Territoires de l'Industrie en Europe (1975–2000)*, eds. Daumas, J.-C., Lamard, P. and Tissot, L. (Presses Universitaires de Franche-Comté, Besançon), pp. 9–34.

Delgado, J. L. (2015). *Del bosque a la fábrica: Técnica y ciencia de la resina de pino en la España contemporánea* (Ph.D. Dissertation, Autonomous University of Madrid).

Delgado, J. L. (2016). Industrialization and Landscape: Gum Resin Pinewoods of France, Spain and United States in the 19th and 20th Centuries, *Estudios Rurales*, 6(11), pp. 48–69.

Desalbres, L. (1929). Industrie de la Terpine et du Terpinéol, *Bulletin de l'Institut du Pin,* 56, pp. 10–14.

Desalbres, L. (1946). La vie financière de l'Union Corporative des Résineux, *Bois et Résineux*, 30 November 1946.

Dessort, P. (1869). Du pin maritime et de ses produits (Dissertation, École supérieure de pharmacie de Paris).

Deville, J. (2009). *L'Incendie meurtrier. Dans la forêt des Landes en août 1949* (Les Éd. des Pompiers de France, Paris).

Dhommée, R. (1927). Ecole pratique d'industrie du pin, *Bulletin de l'Institut du Pin*, 41, pp. 225–232.

Dioscorides (2005). *De Materia Medica* (Olms, Hildesheim).

Dive, H. (1864). *Du Pin Maritime* (Delaroy, Mont-de-Marsan).

Dubourg, J. (1927). Couleurs, Vernis, Peintures, cours de M. le professeur Bourguel, *Bulletin de l'Institut du Pin*, 36, pp. 121–131.

Duffour, A. (1920). Les constituants cristallisés des résines de conifères, *Forêt de Gascogne*, 18.

Duhem, P. (1899). Usines et laboratoires, *Revue Philomathique de Bordeaux et du Sud-Ouest*, 9, pp. 385–400.

Duhem, P. (1901). *Georges Brunel* (Imprimerie G. Delmas, Bordeaux).

Dunwody, R. (1921). Proper methods of distillation and handling in the production of turpentine and rosin, in *Naval Stores: History, Production, Distribution and Consumption,* ed. Gamble, T. (Review Publishing & Printing Company, Savannah), pp. 127–133.

Dupont, G. (1921). L'Institut du Pin: Son but — son organisation — son programme, *Bulletin Officiel de la Direction des recherches scientifiques et industriels et des inventions*, 15, pp. 30–36.

Dupont, G. (1922). *Le Bilan de l'Institut du Pin en 1922* (Feret, Bordeaux).

Dupont, G. (1923). La Forêt Landaise. Vers l'exploitation méthodique de cette richesse nationale. Rôle de l'Institut du Pin, *Revue Philomathique de Bordeaux*, January–February 1923, pp. 22–35.

Dupont, G. (1924). Le pin maritime et les industries dérivées, *Bulletin de l'Institut du Pin*, 1, no page.

Dupont, G. (1924b). *Distillation du Bois* (Gauthier-Villars, Paris).

Dupont, G. (1924c). Le Bulletin de l'Institut du Pin, *Bulletin de l'Institut du Pin*, 1, pp. 1–2.

Dupont, G. (1925). L'opaque dans les Colophanes, *Bulletin de l'Institut du Pin*, 15, pp. 183–185.

Dupont, G. (1926). *Essences de Térébenthine* (Masson et Gauthier-Villars, Paris).

Dupont, G. (1926b). Sur l'avenir économique de la forêt landaise, *Bulletin de l'Institut du Pin*, 25, pp. 413–416.

Dupont, G. (1927). Sur la Coloration des Essences de térébenthine et les moyens pour y remédier, *Bulletin de l'Institut du Pin,* 29, pp. 529–534.

Dupont, G. (1927b). La Récolte et le Traitement de la gemme, *Bulletin de l'Institut du Pin*, 35, pp. 29–35.

Dupont, G. (1927c). Le savon de gemme brute est une invention française, *Bois et Résineux*, 12 June 1927.

Dupont, G. (1927d). Les sous-produits de l'exploitation forestière landaise — leur utilisation rationnelle, *Bulletin de l'Institut du Pin*, 40, pp. 201–204.

Dupont, G. (1928). L'Emploi de la résine dans les savons, *Bulletin de l'Institut du Pin*, 46, p. 56.

Dupont, G. (1929). Rapport sur le fonctionnement et les travaux de l'Institut du Pin en 1928, *Bulletin de l'Institut du Pin*, 60, pp. 149–154.

Dupont, G. (1930). Les Emplois des Goudrons de Bois, *Bulletin de l'Institut du Pin*, 1, pp. 17–24.

Dupont, G. (1931). La Crise des Résineux et la concurrence faite par le Pétrole à l'Essence de Térébenthine, *Bulletin de l'Institut du Pin*, 16, pp. 73–78.

Dupont, G. (1932). Rapport sur le fonctionnement et les travaux de l'Institut du Pin en 1931, *Bulletin de l'Institut du Pin*, 27, pp. 49–58.

Dupont, G. (1933). L'utilisation de l'Essence de térébenthines dans les peintures, *Bulletin de l'Institut du Pin*, 45, p. 176.

Dupont, G. (1933b). Rapport sur le fonctionnement et les travaux de l'Institut du Pin en 1932, *Bulletin de l'Institut du Pin*, 40, pp. 73–77.

Dupont, G. (1935). Maurice Vèzes, *Bulletin de l'Institut du Pin*, 2, p. 24.

Dupont, G. and Barraud, M. (1927). La récolte et le traitement de la gemme. *Bulletin de l'Institut du Pin*, 32, pp. 29–35.

Dupont, G. and Dubaquié, J. (1933). Les résinates et leur emploi comme mouillant et fixatif en agriculture, *Bulletin de l'Institut du Pin*, 44, pp. 146–151.

Dupont, G. and Dubourg, J. (1928). Sur l'acide pyroabiétique, *Bulletin de l'Institut du Pin*, 51, pp. 181–184.

Dupont, G. and Vèzes, M. (1921). Une initiative: L'Institut du Pin, *Sud-Ouest Économique, "La Science dans la Lutte économique moderne"*, 8 December 1921, pp. 1–6.

Dyer, D. and Sicilia, D. B. (1989). From Commodity to Specialty Chemicals: Cellulose Products and Naval Stores at the Hercules Powder Company, 1919–1939, *Business and Economic History*, 18, pp. 59–71.

Earley, L. S. (2004). *Looking for Longleaf. The Fall and Rise of an American Forest* (University of North Carolina Press, Chapel Hill).

Edgerton, D. (2004). The linear model did not exist: Reflections on the history and historiography of science and research in industry in the twentieth century, in *The Science-Industry Nexus: History, Policy, Implications,* eds. Grandin, K. and Wormbs N. (Science History Publications, Sagamore Beach), pp. 31–57.

Egyedi, T. M. (1996). *Shaping Standardization: A Study of Standards Processes and Standards Policies in the Field of Telematic Services* (Delft University Press, Delft).

Ekerholm H. (2012). Cultural meanings of wood gas as automobile fuel in Sweden, 1930–1945, in *Past and Present Energy Societies*, eds. Möllers, N. and Zachmann, K. (transcript-Verlag, Bielefeld), pp. 223–248.

Eldredge, I. F. (1921). How the French turpentine system looked to an American, in *Naval Stores: History, Production, Distribution and Consumption,* ed. Gamble, T. (Review Publishing & Printing Company, Savannah), pp. 167–174.

Feller, R. L. (1964). What's in a Name: Dammar, or, Serendipity in the Library, *The Crucible*, 49, pp. 214–218.

Feng-en Tu (2015). "The Global Life of Camphor and the Making of Synthetic Modern". Workshop on the History of Medicine, Society and the Body at Keio University, 11 March 2015 (unpublished).

Feytaud, J. and Lapparent, P. (1935). Sur l'utilisation des produits du pin comme insecticides et comme mouillants, *Bulletin de l'Institut du Pin,* 6, pp. 105–109.

Finlay, M. R. (1990). The Industrial Utilization of Farm Products and By-products: The USDA Regional Research Laboratories, *Agricultural History*, 64(2), pp. 41–52.

Finlay, M. R. (2004). Old Efforts at New Uses: A Brief History of Chemurgy and the American search for Biobased Materials, *Journal of Industrial Ecology*, 7(3–4), pp. 33–46.

Fortun, M. (2001). Mediated Speculations in the Genomics Futures Markets, *New Genetics and Society*, 20(2), pp. 139–156.

Frank, A. (2014). Aux origines du gemmage, *Sud-Ouest*, 2 October 2014 (www.sudouest.fr/2014/10/02/aux-originesdu-gemmage-1690231-2780.php, last accessed: 11 February 2021).

Frie, F. (1956). *Ruine ou prospérité du gemmage? Recherche d'une solution à la crise des résineux landais* (self published).

Gamble, T. (1921). Early history of the naval stores industry in North America, in *Naval Stores: History, Production, Distribution and Consumption,* ed. Gamble, T. (Review Publishing & Printing Company, Savannah), pp. 17–23.

Gamble, T. (1921b). The world-wide distribution of naval stores, in *Naval Stores: History, Production, Distribution and Consumption,* ed. Gamble, T. (Review Publishing & Printing Company, Savannah), pp. 91–95.

Gamble, T. (1921c). Pages from Wilmington's story as America's first great naval stores port, in *Naval Stores: History, Production, Distribution and Consumption,* ed. Gamble, T. (Review Publishing & Printing Company, Savannah), pp. 31–34.

Gamble, T. (1921d). Charleston's story as a naval stores emporium, in *Naval Stores: History, Production, Distribution and Consumption,* ed. Gamble, T. (Review Publishing & Printing Company, Savannah), pp. 35–36.

Gamble, T. (1921e). Pine trees of the south and their products, in *Naval Stores: History, Production, Distribution and Consumption,* ed. Gamble, T. (Review Publishing & Printing Company, Savannah), pp. 9–16.

Gamble, T. (1921f). Mining for Rosin in the Old North State, in *Naval Stores: History, Production, Distribution and Consumption,* ed. Gamble, T. (Review Publishing & Printing Company, Savannah), pp. 37–39.

Gardner, H. A. and Sward, G. G. (1972). *Paint Testing Manual* (ASTM International, Philadelphia).

Gérard, A. (1931). L'essence de térébenthine dans les vernis et peintures, *Bulletin de l'Institut du Pin,* 13, p. 4.

Gildemeister, E. and Hoffman, F. (1900). *Les Huiles Essentielles* (Schimmel & Cie, Liepzig).

Godin, B. (2006). The Linear Model of Innovation: The Historical Construction of an Analytical Framework, *Science, Technology & Human Values*, 31(6), pp. 639–667.

Gossot, H. and Peuron, R. (2007). Le secteur balistique et spatial Aquitain, *Empreintes: la revue de la Mémoire de Bordeaux, de la Communauté urbaine et de ses communes*, 57.

Grelon, A. (1988). Les écoles d'ingénieurs et la recherche industrielle. Un aperçu historique, *Culture Technique*, 18, pp. 232–238.

Grelon, A. (1996). La naissance de l'enseignement supérieur industriel en France, *Quaderns d'Historia de l'Enginyeria*, 1, pp. 53–81.

Groc (1931). La Genèse et le but de la Fédération des Produits Résineux et du Pin, *Bois et Résineux*, 24 May 1931.

Grochowski, W. (1965). Le Gemmage en Pologne, *Revue Forestière Française*, 6, pp. 432–447.

Harmon, M. A. and Snedeker, R. J. (1997). The archaeological record of tar and pitch production in coastal Carolina, in *Carolina's Historical Landscapes: Archaeological Perspectives*, eds. L.F. Stine, *et al.* (The University of Tennessee Press, Knoxville), pp. 145–160.

Harper, W. (1921). Development of the wood turpentine and wood rosin industry, in *Naval Stores: History, Production, Distribution and Consumption*, ed. Gamble, T. (Review Publishing & Printing Company, Savannah), pp. 245–246.

Hawkins, J. E. (1948). Naval stores research at the University of Florida, in *Naval Stores Review International Yearbook* (Naval Stores Review, New Orleans).

Hazera, L. (1872). Etude du pin maritime et de ses produits (Dissertation, École supérieure de pharmacie de Paris).

Herty, C. (1903). *A New Method of Turpentine Orcharding* (U.S. Department of Agriculture, Bureau of Forestry, Washington).

Hessenland, M. (1935). Harznutzung in deutschen Waldern nach einem neuen Verfahren, *Angewandte Chemie*, 40, p. 636.

Hevly, B. (1996). The Heroic Science of Glacier Motion, *Osiris*, 11, pp. 66–86.

Hicks, E. (1961). *Shellac: Its Origin and Applications* (Chemical Pub. Co., New York).

Holl, J. M. (1997). *Argonne National Laboratory, 1946–96* (University of Illinois Press, Chicago).

Howes, F. N. (1949). *Vegetable gums and resins* (Chronica Botanica Co. Waltham).

Huber, L. and Schlaudt, O. (2015). Introduction, in *Standardization in Measurement* (Routledge, New York).

Huwart, O. (2000). Une filière méconnue de transfert de technologie: l'apport allemand à l'industrie française d'armement après 1945 (Ph.D. dissertation, Institut d'Etudes Politiques, Centre d'histoire de l'Europe du XXe siècle).

Huwart, O. (2002). Missiles et lanceurs spatiaux: La France gaullienne relève un défi technologique et politique (1958–1964), in *Armement et Ve République*, ed. Vaïsse, M. (CNRS Editions, Paris), pp. 375–391.

Jakobs, K. (2010). How People and Stakeholders Shape Standards: The Case of IEEE 802.11, *Lecture Notes in Business Information Processing*, 75, pp. 1–13.

Jas, N. (2000). *Au carrefour de la chimie et de l'agriculture : Les sciences agronomiques en France et en Allemagne, 1840–1914* (Ed. des Archives contemporaines, Paris).

Jaucourt, L. de, (1751). Térébenthine, in *Encyclopédie, ou Dictionnaire Raisonné des Sciences, des Arts et des Métiers*, tome 16 (Chez Samuel Faulche & Cie, Neufchastel).

Johnson, L. and Schaffer, D. (1994). *Oak Ridge National Laboratory: The First Fifty Years* (The University of Tennessee Press, Knoxville).

Jessop, P. (2020). Evidence of a Significant Advance in Green Chemistry, *Green Chemistry*, 22, pp. 13–15.

Kaiserfeld T. (2013). Why New Hybrid Organizations are Formed: Historical Perspectives on Epistemic and Academic Drift, *Minerva*, 51(2), pp. 171–194.

Kohler, R. (2002). *Landscapes and Labscapes: Exploring the Lab-Field Border in Biology* (University of Chicago Press, Chicago).

Kopaczyk, J. M., Warguła, J. and Jelonek, T. (2020). The Variability of Terpenes in Conifers under Developmental and Environmental Stimuli, *Environmental and Experimental Botany*, 180 (online).

Krasnodębski, M. (2014). Interview with Claude Filiatre, former director of the Pine Institute, 10 January 2014.

Krasnodębski, M. (2016). Interview with Claude Filiatre, former director of the Pine Institute, 12 April 2016.

Krasnodębski, M. (2016b). From Distillation to Standardization: A French Perspective on Shaping of Turpentine Spirit (1909–1976), *Ambix*, 63(4), pp. 326–346.

Krasnodębski, M. (2017). Can Science Feed on a Crisis? Expectations, the Pine Institute and the Decline of the French Resin Industry, *Science in Context*, 30(1), pp. 61–87.

Krasnodębski, M. (2018). Beyond Private and Public Research: The Legal and Organizational Reality Behind Industrial Research Institutes in Interwar France, *Minerva*, 56(3), pp. 333–355.

Krasnodebski, M. (2019). Challenging the Pine: Epistemic Underpinnings of Techno-Environmental Inertia, *Journal for the History of Environment and Society*, 4, pp. 41–69.

Krasnodębski, M. (2021). Upscaling Forest Waste: The French Quest for Fuel Autarky after World War I. *Technology and Culture*, 62, pp. 105–127.

Krasnodębski, M. (2022). *Green Chemistry: A Brief Historical Critique* (Peter Lang, Berlin).

Kumar, P. (2012). *Indigo Plantations and Science in Colonial India* (Cambridge University Press, Cambridge).

Lafon (1924). Communication de M. Lafon au 4e Congrès de Chimie Industrielle: Ethers Résiniques, *Bulletin de l'Institut du Pin*, 5, no page.

Langenheim, J. (2003). *Plant Resins: Chemistry, Evolution, Ecology and Ethnobotany* (Timber Press, Portland).

Lascombe, J. (2009). *La Faculté des Sciences de Bordeaux ou L'Université « Bordeaux I » et les pays en voie de développement* (Presses Universitaires de Bordeaux, Bordeaux).

Lemáry, N. (1675). *Cours de chimie, contenant la manière de faire les opérations qui sont en usage dans la médecine, par une méthode facile, avec des raisonnements sur chaque opération, pour l'instruction de ceux qui veulent s'appliquer à cette science* (chez l'auteur, rue Galande, à la Porte Dorée proche la Place Maubert, Paris).

Lescarret, (1957). Essai critique sur l'Administration de la forêt landaise et particulièrement sur le circuit de la gemme, *Bois et Résineux*, 10–20 August 1957.

Lestel, L. (2008). *Itinéraires de chimistes: 1857–2007, 150 ans de chimie en France avec les présidents de la SFC* (EDP Sciences, Paris).

Lockwood, J. E. (1921). Wood rosin and turpentine vs. Gum Rosin and Turpentine, in *Naval Stores: History, Production, Distribution and Consumption,* ed. Gamble, T. (Review Publishing & Printing Company, Savannah), pp. 247–249.

Lombard, R. (1944). Contribution à l'étude des acides résiniques: étude de la composition des gemmes de pin et de la résine d'épicéa. Obtention d'un nouvel acide primaire, l'acide dextrosapinique, *Bulletin de la Société chimique de France*, 11, pp. 201–205.

Lombard, R. (1946). *Produits résineux, gemmes, colophanes & dérivés* (Dunod, Paris).

Lombard, R. (1954). Sur la nature de l'acide dextrosapinique, *Bulletin de la Société chimique de France*, 21, pp. 35–38.

Maison, L. (2013). La fondation et les premiers travaux de l'observatoire astronomique de Bordeaux (1871–1906): histoire d'une réorientation scientifique (Ph.D. dissertation, University of Bordeaux I).

Maison-Soulard, L. and Frigant, V. (2020). *L'industrie aérospatiale en Nouvelle-Aquitaine: un siècle d'histoire et de patrimoine* (Éditions Le Festin, Bordeaux).

Maizeret, C. (2005). *Les Landes de Gascogne* (Delachaux et Niestlé, Paris).

Manuel J. T. (2021). Lessons from a Forgotten Fuel: Assessing the Long History of Alcohol Fuel Advocacy and Use in the United States, *History and Technology*, 37, pp. 411–428.

Marcard, R. (1936). L'emploi obligatoire de la résine en savonnerie, *Bulletin de l'Institut du Pin*, 20, p. 169–172.

Marcy Ch. (1941). Editorial, *Bois et Résineux*, 19 October 1941.

Maydieu, P. (1931). Les emplois et le Commerce de la Colophane, *Bulletin de l'Institut du Pin*, 19, pp. 145–149.

Maydieu, P. (1931b). Les usages des colophanes, *Bois et Résineux*, 31 May 1931.

Mercelis, J. (2020). *Beyond Bakelite: Leo Baekeland and the Business of Science and Invention* (MIT Press, Cambridge).

Merklen, F. (1928). Le projet de loi sur la Répression des Fraudes dans le Commerce des Savons, *Bois et Résineux*, 5 August 1928.

Merklen, F. (1934). Raisons de l'emploi de la résine dans les savons, *Bulletin de l'Institut du Pin*, 59–60, pp. 227–231.

Mims, L. W. (1949). D.D. naval stores industry has seen interesting changes, in *Naval Stores Review International Yearbook* (Naval Stores Review, New Orleans).

Moral, C. del. (1921). History of the naval stores in Spain, in *Naval Stores: History, Production, Distribution and Consumption,* ed. Gamble, T. (Review Publishing & Printing Company, Savannah).

Moriniaux, V. (1999). Les Français face à l'enrésinement: XVIe-XXe siècles (Ph.D. dissertation, University Paris-Sorbonne).

Morris, P. (1995). *The Matter Factory* (Reaktion Books, London).

N. N. (1919). *Hearings before the Committee on Agriculture and Forestry United States Senate, Sixty-Sixth Congress on S. 1351 Bill Establishing Standard Grades of Naval Stores, Preventing Deception in Transactions in Naval Stores, Regulating Traffic Therein, and for Other Purposes* (Government Printing Office, Washington).

N. N. (1924). L'emploi de la colophane en savonnerie, *Bulletin de l'Institut du Pin*, 4, no page.

N. N. (1924b). Rapport de la Commission désignée par la société des experts chimistes pour étudier la question de l'essence de térébenthine, *Bulletin de l'Institut du Pin*, 1, no page.

N. N. (1924c). Procédé de préparation des résinates métalliques neutres, *Bulletin de l'Institut du Pin*, 4, no page.

N. N. (1924d). Analyses, *Bulletin de l'Institut du Pin*, 1, no page.

N. N. (1925). Rapport sur les travaux de l'Institut du Pin en 1923–1924, *Bulletin de l'Institut du Pin*, 8, pp. 1–8.

N. N. (1925b). L'Institut du Pin d'Amérique, *Bulletin de l'Institut du Pin*, 9, p. 27.

N. N. (1927). Editorial. *Bois et Résineux*, 6 February 1927.

N. N. (1927b). Editorial, *Bois et Résineux*, 20 February 1927.

N. N. (1927c). Rapport sur le fonctionnement et les travaux de l'Institut du Pin en 1926, *Bulletin de l'Institut du Pin*, 35, pp. 73–77.

N. N. (1927d). Le cycle des résineux, *Bois et Résineux*, 9 January 1927.

N. N. (1928). Rapport sur le fonctionnement et les travaux de l'Institut du Pin en 1927, *Bulletin de l'Institut du Pin*, 46, pp. 49–55.

N. N. (1928b). Les resineux dans une impasse, *Bois et Résineux*, 30 March 1930.

N. N. (1930). Une victoire pour les résineux, *Bois et Résineux*, 30 March 1930

N. N. (1930b). L'opinion d'un propriétaire, *Bois et Résineux*, 7 December 1930.

N. N. (1931). Les grades de la colophane, *Bulletin de l'Institut du Pin*, 18, 1931, p. 125.

N. N. (1931b). Constitution de la Matière et Valence Chimique, *Bulletin de l'Institut du Pin*, 19, pp. 167–169.

N. N. (1932). Editorial, *Bois et Résineux*, 15 June 1932.

N. N. (1932b). Loi, décret et circulaire sur la répression des fraudes dans le commerce de l'Essence de térébenthine et des Produits résineux, *Bulletin de l'Institut du Pin*, 35, pp. 251–257.

N. N. (1932c). Congrès de la Forêt et de ses Industries, discussion sur le rapport de Marcelle Barraud, *Bulletin de l'Institut du Pin*, 36, pp. 265–270.

N. N. (1933). Compte rendu de l'Assemblée générale de l'Institut du Pin, *Bois et Résineux*, 2 April 1933.

N. N. (1933b). Manifestation de sympathie à l'adresse de M.G Dupont, *Bulletin de l'Institut du Pin*, 47, pp. 20.

N. N. (1933c). Advertisements, *Bulletin de l'Institut du Pin*, n. 40, 1933, pp. 30.

N. N. (1934). Editorial, *Bois et Résineux*, 7 January 1934.

N. N. (1939). Rapport sur l'utilisation de la résine dans les liants routiers, *Bois et Résineux*, 9 July 1939.

N. N. (1939b). Le résultat de deux mois de fonctionnement de l'Union corporative des résineux, *Bois et Résineux*, 30 July 1939.

N. N. (1939c). Vers la centrale unique de vente des résineux, *Bois et Résineux*, 2 April 1939.

N. N. (1940). Ce qu'est l'U.C.R., *Bois et Résineux*, 3 March 1940.

N. N. (1940b). L'organisation de la production résinière, *Bois et Résineux*, 22 December 1940.

N. N. (1942). L'Institut du pin: seul laboratoire officiel, *Bois et Résineux*, 6 December 1942.

N. N. (1943). Editorial, *Bois et Résineux*, 31 January 1943.

N. N. (1947). Note in *Naval Stores Review*, 5 April 1947, p. 3.

N. N. (1948). Étude sur la possibilité de modernisation du gemmage, *Bois et Résineux*, 29 February 1948.

N. N. (1949). L'U.C.R. dans la vie économique et sociale, *Bois et Résineux*, 20 November 1949.

N. N. (1949b). L'U.C.R. et l'Institut du Pin. *Bois et Résineux*, 20 January 1949.

N. N. (1952). Activation de la production de la gemme, *Bois et Résineux*, 29 February 1952.

N. N. (1953). Les résineux au salon de la chimie, *Bois et Résineux*, 15 July 1953.

N. N. (1958). Rapport moral du Conseil d'Administration de la N.U.C.R, *Bois et Résineux*, 10 January 1958.

N. N. (1959). L'emploi de la térébenthine dans la propulsion des fusées, *Bois et Résineux*, 30 November 1959.

N. N. (1962). Editorial, *Chimie et Industrie*, 87(4), p. 450.

N. N. (1963). Les Résineux seront-ils sauvés?, *Bois et Résineux*, 10 October 1963.

N. N. (1964). Rapport moral du Conseil d'Administration de la N.U.C.R, *Bois et Résineux*, 20 January 1964.

N. N. (1998). Editorial, *Naval Stores Review*, 108(1), p. 1.

Nicke, W. and DeMert, L. (1942). De woods of pine: a song dedicated to the Pine Institute of America, *Forest Facts*, 14.

Nieddu, M., Vivien, F.-D., Garnier, E. and Bliard, C. (2014). Existe-t-il réellement un nouveau paradigme de la chimie verte?, *Natures Sciences Sociétés,* 22, pp. 103–113.

Nobel Prize Committee (2005). Press Release (www.nobelprize.org/prizes/chemistry/2005/press-release/, last accessed 17 February 2023).

Nye, M. J. (1986). *Science in the Provinces: Scientific Communities and Provincial Leadership in France, 1860–1930* (University of California Press: Los Angeles).

Nye, M. J. (1993). *From Chemical Philosophy to Theoretical Chemistry: Dynamics of Matter and Dynamics Of Discipline, 1800–1950* (University of California Press, Berkeley).

Ortal, Lafon, Guirauton (1935). Incorporation des produits du pin maritime dans les goudrons et bitumes à l'usage de revêtements routiers, *Bulletin de l'Institut du Pin*, 3, pp. 54–56.

Outland, R. (2004). *Tapping the Pines* (Louisiana State University Press, Baton Rouge).

Panda H. (2008). *Handbook on Oleoresin and Pine Chemicals (Rosin, Terpene Derivatives, Tall Oil, Resin & Dimer Acids)* (Asia Pacific Business Press, Delhi).

Pasteur, L. (1927). Science et patrie? in *Pasteur, lettres et discours*, ed. Radot R. V. (Les Bibliophiles Comtois, Besançon), pp. 179–184.

Pasquet, L. (1930). Rapport de M. Louis Pasquet: l'Emploi de la colophane dans les savons doit être libre, *Bois et Résineux*, 6 June 1930.

Patton, E. L. and McKennon, F. L. (1950). Research in the naval stores industry, in *Naval Stores International Yearbook* (Naval Stores Review, New Orleans).

Paul, H. (1980). Apollo courts the Vulcans: The applied science institutes in nineteenth-century French science faculties, in *The Organization of Science and Technology in France 1808–1914*, eds. Fox R. and Weisz G. (Cambridge University Press, Cambridge), pp. 151–181.

Paul, H. W. (1985). *From Knowledge to Power: The Rise of the Science Empire in France, 1860–1939* (Cambridge University Press, Cambridge).

Paul, H. W. (1996). *Science, Vine and Wine in Modern France* (Cambridge University Press: Cambridge).

Penrose, E. T. and Pitelis, C. (2002). *The Growth of the Firm: The Legacy of Edith Penrose* (Oxford University Press, New York).

Pestre, D. (1990). Louis Néel et le magnétisme à Grenoble. Récit de la création d'un empire dans la province française 1940–1965, *Cahiers pour l'histoire du CNRS*, 8.

Pestre, D. (1992). *Physique et physiciens en France, 1918–1940* (Edition des Archives contemporaines, Paris).

Petit, A. (2013). Histoire du concept d'ion au XIXe siècle (Ph.D. dissertation, Nantes University).

Poincaré, H. (1904). A Short Contribution in L'Elite intellectuelle et la démocratie, *Revue Bleue*, 23, p. 708.

Poisson, A. (1929). Les Savons résineux, *Bulletin de l'Institut du Pin*, 56, p. 41.

Porter, R. S. (1978). Gentleman and Geology: The Emergence of a Scientific Career, 1660–1920, *Historical Journal*, 21(4), pp. 809–836.

Portmann (1932). Le savon de résine, *Bois et Résineux*, 14 August 1932.

Posedel, J. and Lawton, S. (2011). *Hercules* (Arcadia Publishing, Charleston).

Poudenx, E. (1947). Le Gemmage au Portugal, *Bois et Résineux*, 10 October 1947.

Prat, Ch. (1947). Conférence faite à la faculté des Sciences de Bordeaux, le 1er Juillet 1947, *Bois et Résineux*, 10 October 1947.

Prat, Ch. (1947b). Défense, aménagement et remise en valeur de la forêt de Gascogne, Rapport présenté au Congrès des Gemmeurs des 8 et 9 novembre à Sabres, *Bois et Résineux*, 30 November 1947.

Prizer, T. C. (2009). Pining for turpentine: Critical nostalgia, memory, and commemorative expression in the wake of industrial decline (Master dissertation, University of North Carolina).

Ranga, M. and Etzkowitz, H. (2013). Triple Helix Systems: An Analytical Framework for Innovation Policy and Practice in the Knowledge Society, *Industry and Higher Education*, 27(4), pp. 237–262.

Reed, G. (1995). *Crusading for Chemistry: The Professional Career of Charles Holmes Herty* (University of Georgia Press, Athens).

Reinhardt, C. (2001). Disciplines, research fields and their boundaries, in *Chemical Sciences in the 20th Century: Bridging Boundaries,* eds. Hoffmann R. and Reinhardt C. (Wiley-VCS, Weinheim), pp. 1–13.

Roberts, J. (2006). Creating Green Chemistry: Discursive Strategies of a Scientific Movement (Ph.D. Dissertation, Faculty of Virginia Polytechnic Institute and State University).

Rouin, G. (1924). L'Essence de térébenthine et ses dérivés, *Bulletin de l'Institut du Pin*, 2, no page.

Rouin, G. (1928). Contribution à l'Etude des Acides Résiniques, *Bulletin de l'Institut du Pin*, 49, 1928, pp. 121–131.

Rusterholz, R. (1958). Rapport moral du Conseil d'Administration de la N.U.C.R [1957–1958], *Bois et Résineux*, 20 August 1958.

Rusterholz, R. (1961). Rapport moral du Conseil d'Administration de la N.U.C.R., *Bois et Résineux*, 10 January 1961.

Rusterholz, R. (1962). Rapport moral du Conseil d'Administration de la N.U.C.R [1961–1962], *Bois et Résineux*, 20–30 December 1962.

Rusterholz, R. (1964). Rapport moral du Conseil d'Administration de la N.U.C.R [1962–1963], *Bois et Résineux*, 20 January 1964.

Ružička, L. (1929). Sur la constitution des acides abiétiques, lévo et dextro-pimariques, *Bulletin de l'Institut du Pin*, 59, pp. 112–122.

Samanos, E. (1864). *Traité de la culture du pin maritime: comprenant des études sur la création des forêts, leur entretien, leur exploitation, et la distillation des produits résineux* (Librairie agricole de la Maison rustique, Paris).

Sargos, J. (1997). *Histoire de la Forêt Landaise: du désert à l'âge d'or* (L'Horizon chimérique, Bordeaux).

Sargos, R. (1931). Le procédé de gemmage Bellini delle Stelle, *Bulletin de l'Institut du Pin*, 19, pp. 152–156.

Sargos, R. (1931b). La Crise Résinière peut-elle être résolue par une amélioration du Gemmage?, *Actes et comptes rendus de l'Association des Techniciens de l'Agriculture du S.-O.* 3.

Sargos, R. (1938). Statistique de production. Bois et gemme de la Forêt Landaise — 1937, *Bulletin de l'Institut du Pin*, 39, pp. 49–63.

Sargos, R. (1949). *Contribution à l'histoire du boisement des Landes de Gascogne* (Editions Delmas, Bordeaux).

Schnurrenberger, E. J. (1939). Le procédé de saccharification du bois de Scholler-Tornesch, *Bulletin de l'Institut du Pin*, 51–52, pp. 56–74.

Scofield, F. (1941). *Color Specifications for Transparent Materials, Symposium on Color* (American Society for Testing Materials, Philadelphia).

Seitz, F. (2007). *A Selection of Highlights from the History of the National Academy of Sciences, 1863–2005* (University Press of America, New York).

Selin, C. (2007). Expectations and the Emergence of Nanotechnology. *Science Technology Human Values*, 32(2), pp. 196–220.

Sell, J. T. and Foussereau, J. (1989). Contact Allergy to Manilla Resin. Nomenclature and Physico-chemistry of Manilla, Kauri, Damar and Copal Resins, *Contact Dermatitis*, 21(4), pp. 228–238.

Servos, J. (1993). Research Schools and Their Histories, *Osiris*, 8, pp. 3–15.

Shapin, S. (2010). *Never Pure* (John Hopkins University Press: Baltimore).

Shingler, G. P. (1948). Recent achievements in naval stores research by the Naval Stores Division, Bureau of Agriculture and Industrial Chemistry, in *Naval Stores Review International Yearbook* (Naval Stores Review, New Orleans).

Shinn, T. (2002). Nouvelle production du savoir et triple hélice. Tendances du prêt-à-penser les sciences, *Actes de la recherche en sciences sociales*, 141(1), pp. 21–30.

Smith, J. G. (1979). *The Origins and Early Development of the Heavy Chemical Industry in France* (Oxford University Press, Oxford).

Snitter, P. (1933). Viellissement des vernis à base d'huile, *Bulletin de l'Institut du Pin*, 47, pp. 223–252.

Teichmann, H. (2011). International Standardization of Technical and Scientific Terminology, *Homo Oeconomicus*, 27, pp. 527–541.

Teissier, P. (2014). *Une histoire de la chimie du solide. Synthèses, formes, Identités* (Hermann, Paris).

Timmermans, S. and Epstein, S. (2010). A World of Standards but not a Standard World: Toward a Sociology of Standards and Standardization, *Annual Review of Sociology*, 36, pp. 69–89.

Tingry, P. F. (1803). *Traité théorique et pratique sur l'art de faire et d'appliquer les vernis* (G. J. Manget Librairie, Genève).

Tinniswood, A. (2019). *The Royal Society* (Head of Zeus Ltd, London).

Tool, M. R. (1993). *Institutional Economics: Theory, Method, Policy* (Kluwer, Dordrecht).

Torok, B. and Dransfield, T. (2017). *Green Chemistry: An Inclusive Approach* (Elsevier, Amsterdam).

Trouis, E. (1927). Les routes spéciales réservées aux automobiles, *Bulletin de l'Institut du Pin*, 32, pp. 1–5.

Tschirch, A. (1900). *Die Harze und die Harzebehalter* (Gebr. Borntraeger, Lepizig).

Tschirch, A. (1926). Résines, Baumes, Caouthouc et Guttapercha, *Bulletin de l'Institut du Pin*, 28, pp. 507–514.

Tze Chiang, *et al.*, (1968). *A Study of the Problems and Potentials of the Gum Naval Stores Industry* (Georgia Institute of Technology, Atlanta).

Uekötter, F. (2021). The Revolt of the Chemists: Biofuels, Agricultural Overproduction, and the Chemurgy Movement in New Deal America, *History and Technology*, 37, pp. 429–445.

Varnoteaux, Ph. (2002). *Les origines et les enjeux de la conquête de l'espace en France, 1944–1962: des héritages aux premières structures et études*, vol. 1 (Presses universitaires du Septentrion, Villeneuve d'Ascq).

Varnoteaux, Ph. (2003). La naissance de la politique spatiale française, *Vingtième Siècle. Revue d'histoire*, 77(1), p. 59–68.

Veitch, F. P. (1921). What Uncle Sam does for the naval stores industry, in *Naval Stores: History, Production, Distribution and Consumption*, ed. Gamble, T. (Review Publishing & Printing Company, Savannah), pp. 135–138.

Veitch, F. P. (1924). The Federal Naval Stores Act, *Journal of Industrial & Engineering Chemistry*, 16(6), p. 640.

Veitch, F. P. (1928). Naval Stores Research in the Bureau of Chemistry and Soils, *Naval Stores Review*, 38(26), 29 September 1928.

Veitch, F. P. (1934). The Naval Stores Station of the Bureau of Chemistry and Soils, *USDA Miscellaneous Publications*, 206.

Vèzes, M. (1902). *Revue des progrès réalisés dans l'étude chimique de la colophane* (Impr. scientifique et littéraire Bussière, Bordeaux).

Vèzes, M. (1903). *Rapport sur le fonctionnement du laboratoire de chimie appliquée à l'industrie des résines dans l'année 1902–1903* (Imprimerie G. Gounouilhou, Bodeaux).

Vèzes, M. (1905). *Rapport du fonctionnement du Laboratoire de chimie appliquée à l'industrie des résines pendant l'année universitaire 1903–1904* (Imprimerie G. Gounouilhou, Bordeaux).

Vèzes, M. (1908). *Rapport du fonctionnement du Laboratoire de chimie appliquée à l'industrie des résines pendant l'année universitaire 1906–1907* (Imprimerie G. Gounouilhou, Bordeaux).

Vèzes, M. (1922). *Revue des travaux récents sur la chimie des résines*, 1900–1921 (Impr. Scientifique et littéraire Bussière, Bordeaux)

Vèzes, M. (1932). Institut du Pin, *Bois et Résineux*, 25 June 1932.

Vèzes, M. (1934). Une Fondation Régionale. L'Institut du Pin de l'Université de Bordeaux, *Revue Internationale de l'Enseignement*, 1, 15 January 1934.

Vèzes, M. and Dupont, G. (1924). *Résines et Térébenthines* (J.-B. Baillière et fils, Paris).

Vèzes, M. and Mouline, J. (1901). Sur l'essai technique de l'essence de térébenthine des Landes (III): Dosage acidimétrique de la colophane, *Procès-verbaux de la Société des sciences physiques et naturelles de Bordeaux*, 2 May 1901.

Vèzes, M. and Mouline, J. (1901b). Sur l'essai technique de l'essence de térébenthine des Landes (V): Dosage réfractométrique des adultérants anormaux peu volatils, *Procès-verbaux de la Société des sciences physiques et naturelles de Bordeaux*, 23 May 1901.

Vèzes, M. and Mouline, J. (1902). Sur l'essai technique de l'essence de térébenthine des Landes (X): Etude des adultérants anormaux très volatils, *Procès-Verbaux de la Société des Sciences Physiques et Naturelles de Bordeaux*, 12 June 1902.

Vie, G. (1943). Lubrifiants de Synthèse à partir du Cracking des Colophanes, *Bois et Résineux*, 4 April 1943.

Voillequin, B. (2010). *La catalyse en France (1944–2004): Dynamiques disciplinaires et régimes de production de savoir* (Editions universitaires européennes, Paris).

Warren, F. A. (1958). *Rocket Propellants* (Reinhold Publishing Corporation, New York).

Weber, G. A. (1928). *The Food, Drug and Insecticide Administration: Its History, Activities and Organization* (Johns Hopkins Press, Baltimore).

Wernicke, O. H. L. (1925). Proposed Plan of Naval Stores Institute, *Proceedings of the Seventh Southern Forestry Congress* (Seeham Printery, Durham), pp. 143–152.

Wilks, E. (2001). *Industrial Polymers Handbook: Products, Processes, Applications* (Wiley VCH, New York).

Woodhouse E. J. and Breyman, S. (2005). Green Chemistry as Social Movement? *Science, Technology, & Human Values*, 30, pp. 199–222.

Woodroof, M. and Patton, E. L. (1956). Utilizaton research on gum naval stores, in *Naval Stores Review International Yearbook* (Naval Stores Review, New Orleans).

Woolsey, T. S. (1921). Conservative turpentining by the French: The methods pursued in France, Algiers and Corsica, in *Naval Stores History, Production, Distribution and Consumption*, ed. Gamble, T. (Review Publishing & Printing Company, Savannah), pp. 175–188.

Zinkel, D. F. (1989). *Naval Stores, Production, Chemistry, Utilization* (Pulp Chemical Association, New York).

INDEX OF PEOPLE

INDEX OF ORGANISATIONS

www.ingramcontent.com/pod-product-compliance
Lightning Source LLC
LaVergne TN
LVHW010330220926
848484LV00004BB/868
9789811252853